D0433320

First published 2008 AD
This edition

Published by Wooden Books Ltd.
8A Market Place, Glastonbury, Somerset

British Library Cataloguing in Publication Data
Cheshire, G.
Evolution

A CIP catalogue record for this book is
available from the British Library

ISBN 978 1904263 80 7

Recycled paper throughout.

Printed and bound in England by
The Cromwell Press, Trowbridge, Wiltshire.
100% recycled papers made specially for
Wooden Books by Paperback.

EVOLUTION

A LITTLE HISTORY OF A GREAT IDEA

Gerard Cheshire

To all who respect life ... in all its many forms

Thanks to Peter Spring for edits and suggestions throughout, to William Spring for graphics on pages 13, 15, 19, 51, 53, 55 and 57, to Chris Taylor for drawings on pages 17, 23 and 47, to Dan Goodfellow for illustrations on pages 11 and 45, to Matt Tweed for the cartoon on page 43 and to John Martineau at Wooden Books for supervising, editing and designing this book.

Other recommended reading: Blackmore, S. The Meme Machine. *Oxford University Press (1999). Conway Morris, S.* Life's Solution, Inevitable Humans in a Lonely Universe. *Cambridge University Press (2003). Dawkins, R.* The Selfish Gene. *Oxford University Press (1976, 1992). Dawkins, R.* The Blind Watchmaker. *Longman Scientific & Technical Ltd (1986). Distin, K.* The Selfish Meme. A Critical Reassessment. *Cambridge University Press (2005). Darwin, C., (Wilson, E. O. ed).* From So Simple A Beginning. The Four Great Books of Charles Darwin. *W.W. Norton & Company Ltd. (2006). Gardner, J.* Biocosm, *Inner Ocean (2003). Gee, H.* Deep Time: Cladistics. The Revolution in Evolution. *Fourth Estate (2000). Mayr. E.* What Evolution Is. *Phoenix; Orion Books Ltd (2002). Rees, M.* Just Six Numbers, *Phoenix (2000). Ridley, M.* Genome. *Fourth Estate (1999). Ridley, M.* The Red Queen, Sex and the Evolution of Human Nature, *Penguin (1994).*

Above: The great hope of many creationists is to find a human skull dated to an epoch which disproves the Darwinian theory of evolution. Not a single shred of evidence, however, has yet emerged from the fossil record which disproves Darwin's basic theory.

CONTENTS

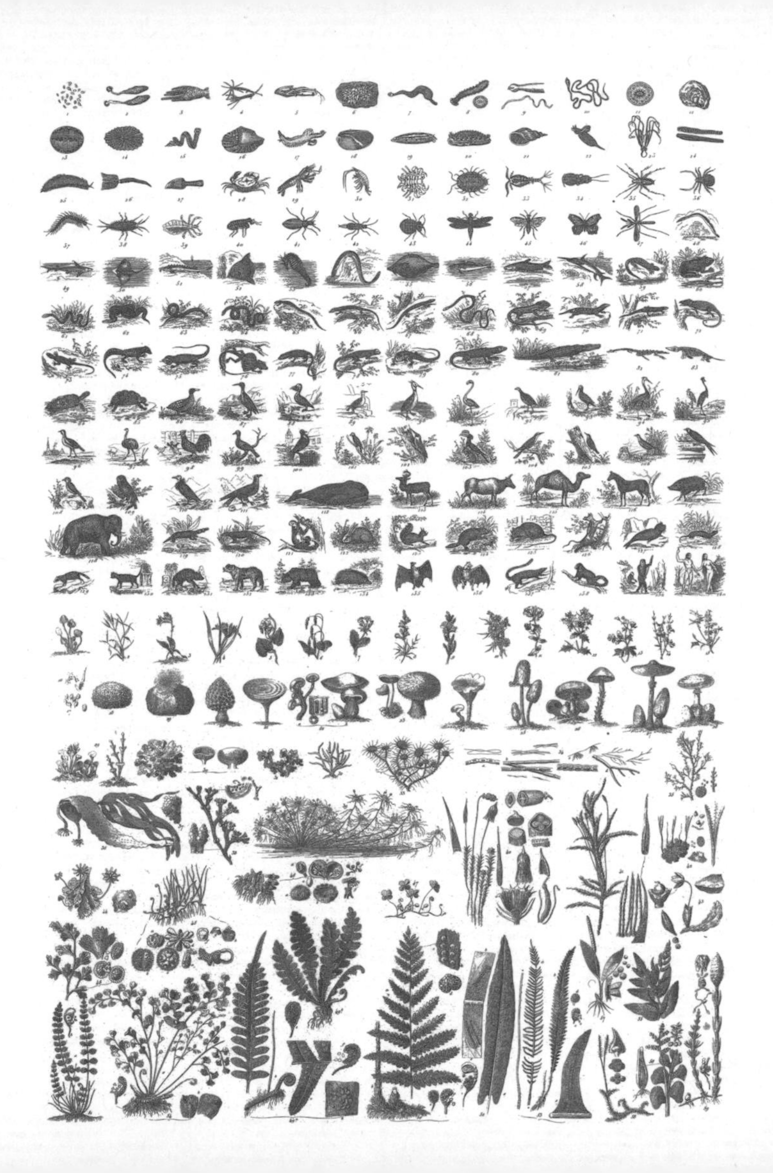

INTRODUCTION

There are few peoples on Earth who do not have a creation myth. The Native American Iroquois believe the world and everything on it was created by Sky People, the ancient Japanese believed the world was created by gods who grew from a single green shoot and many people today still fervently believe that the universe was created in one form or another by a god.

This little book tells the remarkable tale of a modern creation story that has been meticulously pieced together over the last century and a half by hundreds of thousands of botanists, zoologists, chemists and biologists working all across the world. Instead of the rich symbolism of myth or the rote certitudes of religious doctrine, it is generally couched in the difficult language of experimental science. Still as terrifying to as many people today as it was when first publicised by Charles Darwin in 1859, it tells the unlikely tale of a bacterium that became a kind of worm, that became a fish, that became a reptile, that became a sort of rodent, who became an ape, who became a human, who left Africa, and became you.

Like many creation myths it sounds fantastic. Like all good stories it is full of sex, death, family struggles, kindness and friendship. It is a story some people have only just heard for the first time, others never before, because we are only now filling in the details. And yet the story is not finished at all, it is still unfolding. If we survive the era of mass extinctions we are manufacturing for our fellow travelers on this little earth-crusted ball of fire, we will become other things yet.

Life's Great Family

the fog rises

Amidst the foggy ideas of past centuries there arose occasional glimpses of a strange new notion: that humanity, along with all other organisms, instead of being created outright, had instead arisen through a process of biological adaptation—*evolution*.

Publishing *Systema Naturae* in 1735, Carl Linnaeus (1707-78) replaced the classical categorization of animals by their mode of movement, with the system of *kingdoms*, *phyla*, *classes*, *orders*, *families*, *genera* and *species* still used today. It seemed evident that these families of animals and plants had evolved in some way from common ancestors, or from one another, and by the 1800s scientists were trying to work out exactly how. In 1809 Jean-Baptiste Lamarck (1744-1829) proposed that species evolved via acquired characteristics, so that subtle (and often useful) changes made to their design during their own lifetimes (e.g., a tennis player's better developed arm muscles) were passed on to their offspring. This theory, however, although popular, had serious flaws. It turned out that offspring often varied wildly from their parents, and, importantly, that characteristics acquired over a lifetime, such as injuries or larger muscles, could not be passed down the generations either.

The theory was not working. Something was missing.

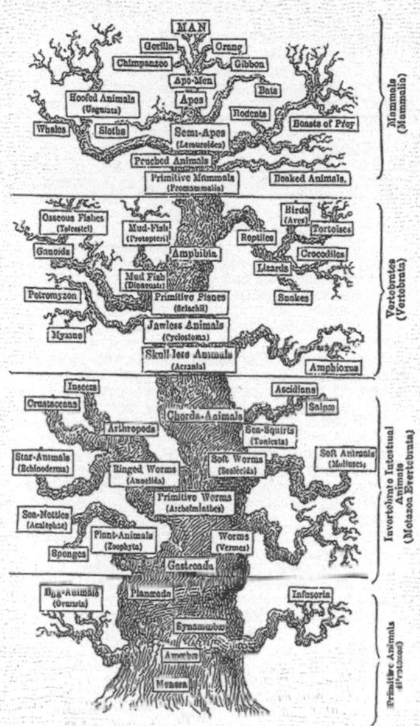

Left: An early Linnaean tree of life, the kingdoms shown as a hierarchy with mammals at the top, and man at the summit. Although a breakthrough, these early versions were not so very far from the medieval Chain of Being, the hierachy of souls, with God at the top, then angels, humans, animals, vegetables and finally minerals, each kingdom having natural authority over and control of those beneath it.

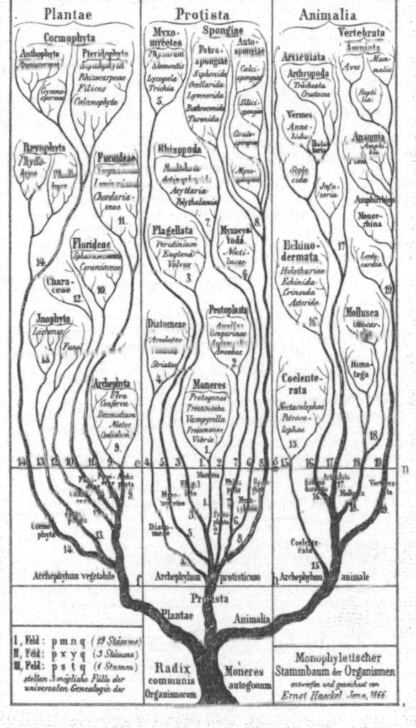

Right: Ernst Haeckel's original tree of life, drawn in 1866, dividing living things into three basic groups, plants, animals and protists (a diverse group of eukaryotes, or multicellular organisms, which do not fit into the other eukaryotic kingdoms). Haeckel coined the term "protist" for this diagram. Modern classifications vary from this diagram in certain important ways (e.g. fungi are today considered a kingdom of their own). To see the modern versions of the tree of life, turn to the end pages of this book *(pages 50-58).*

The Great Idea

eat, breed, adapt and pass it on

In 1859, after over 25 years collecting specimens and studying variations between species, particularly barnacles, Charles Darwin (1809-82) revealed his theory of evolution by natural selection. It directly contrasted with Lamarck's theory. The fact that offspring vary in their characteristics, suggested Darwin, was enough to allow nature to select in favor of those individuals slightly better suited to perpetually changing environments. Tiny changes, each conferring small advantages, could mount up, over many generations, to produce large differences, even new species. In 1864 Herbert Spencer (1820-1903) coined the phrase 'survival of the fittest,' attempting to encapsulate this idea.

Darwinism usurped Lamarckism, though no one, including Darwin, was yet able to furnish an empirically evidenced explanation for the mechanism that promoted the variation that allowed natural selection to work. Nevertheless, *gemmules*, his descriptive term for the particles responsible for biological transfer, were, unknown to him at the time, in many respects similar to Mendel's pairs (*page 8*).

Darwin's theory also suggested that man had descended from ancestral apes. A revolutionary concept at the time, his theory of evolution questioned humanity's place in the universe and challenged long-held beliefs about the nature of creation.

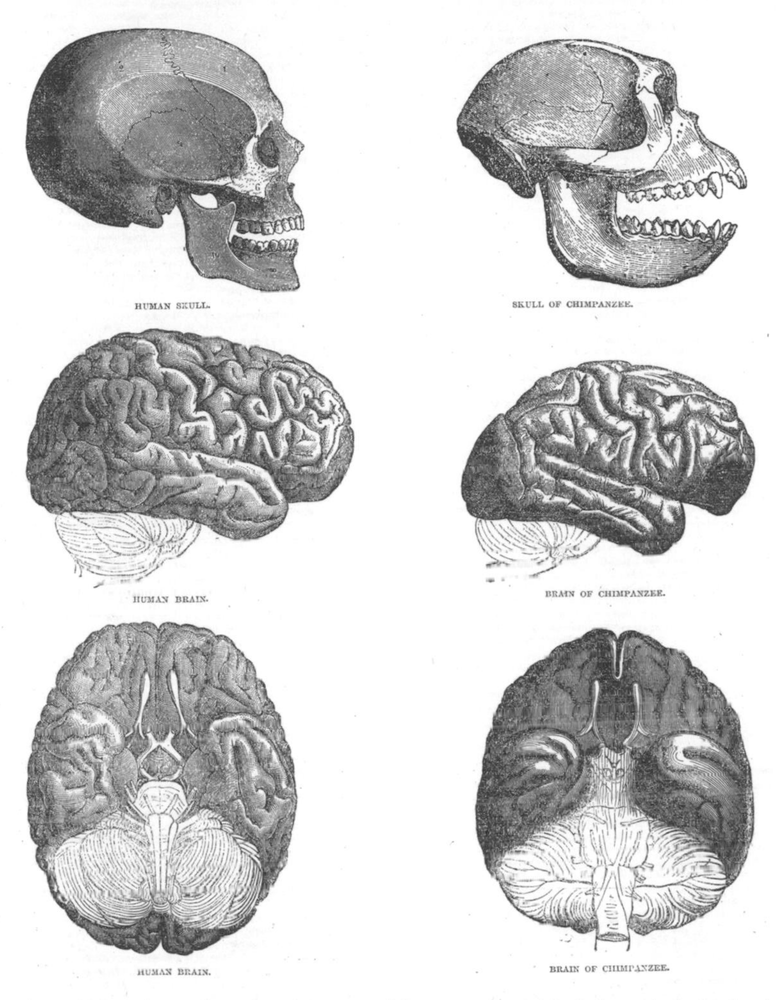

Above and left: Striking similarities between humans and chimpanzees suggested that both were closely related, and that man was a species of ape. No piece of evidence to the contrary has ever been found since, only further support.

Living Proof
and dead ends

In support of his evolutionary theory Darwin found important examples of evolution in action. One was *artificial selection*, or *eugenics*. Darwin argued that humans had created domesticated plants and animals by imposing selective processes on captive populations. Careful breeding, he suggested, created desirable characteristics in dogs, cats, horses, pigeons and chickens (*opposite top*) in much the same way that nature did in the wild.

Traveling on the *Beagle* (1831-36), Darwin had noticed groups of closely related species that appeared to have adapted to slightly different environmental demands. Examining the exotic reptiles and birds of the Galapagos archipelago in 1835, he found that each island had its own idiosyncratic species of tortoise (*opposite*) and finch (*below*), demonstrating that isolation events had allowed natural selection to take populations in different evolutionary directions on different islands from a common ancestor.

Two problems remained: Firstly, Darwin had only demonstrated *lateral* evolution, not *longitudinal* evolution; species could adapt and vary on a theme, but a tortoise appeared to remain a tortoise and a bird a bird, so his theory didn't yet explain how wholly new types of animal and plant came about. Secondly, he lacked a provable mechanism behind this variation.

Above: Selective human breeding of cattle has created many hundreds of breeds, each with specific qualities. Some are bred for milk, others for meat, some for hot climates, others for freezing hillsides. Chickens too are selectively bred, some for eggs, others for meat. Darwin bred pigeons at his home to better understand the process of artificial selection and get a first-hand grasp of how quickly small variations in individuals could be passed into entire populations.

Left: Giant tortoises on the Galapagos Islands. There are 11 existing species of tortoise spread over the islands, all probably descended from a single ancestor. On dry islands, where little grows except cacti, taller tortoises with longer necks had the advantage over their shorter counterparts. Taller cacti also tended to survive being eaten and are found on these islands, an example of an evolutionary arms race. The tortoises allow the finches on the islands (far left) to peck ticks from their skin, which suits both parties well, and is an example of a mutualistic symbiotic relationship.

The Unsung Monk

peas and their peculiar traits

While Darwin was pondering over his mechanism, a Moravian monk, Gregor Mendel (1822-84), had already been experimenting with heredity for years. Starting in 1856, Mendel had begun breeding pea plants on a hunch that inheritance was mathematically predictable. By 1865 he had tested over 29,000 plants and amassed enough evidence to show that it was possible to accurately forecast the ratios between paired traits (such as smooth vs wrinkled peas, or tall vs dwarf plants) after controlled crossbreeding. For example, crossing purebred tall and dwarf pea plants produced only tall specimens. However, crossing these with each other again, dwarfness reappeared in the next generation, with tall:dwarf plants in a 3:1 ratio. Mendel concluded that pairs of particles (now known as *alleles*), one *dominant* and one *recessive*, were at work (*see example, opposite top*).

Mendel was right. We also now know that other species, such as snapdragon flowers, can also display *incomplete dominance* when red and white varieties are crossed (*lower example opposite*). And there is *codominance*, where neither allele is recessive. An example is the ABO blood-group system, which is controlled by three alleles, *A*, *B* and *o*. *o* is recessive to both *A* and *B*, and causes O-type blood, while *A* and *B* are codominant. You inherit two alleles, one from each parent, and can therefore either end up with A(*AA*, *Ao*), B(*BB*, *Bo*), AB(*AB*), or O(*oo*) blood groups. We also now know that just one letter change on chromosome 9 makes the difference between O and A, but more of that to come.

Darwin didn't hear of Mendel's work and it was not fully recognised until 1900 by William Bateson (1861-1926).

DOMINANT *and* RECESSIVE TRAITS

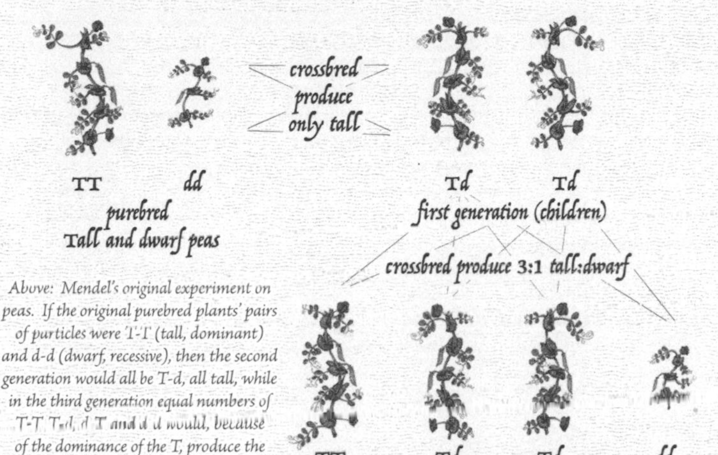

Above: Mendel's original experiment on peas. If the original purebred plants' pairs of particles were T-T (tall, dominant) and d-d (dwarf, recessive), then the second generation would all be T-d, all tall, while in the third generation equal numbers of T-T, T-d, d-T and d-d would, because of the dominance of the T, produce the observed 3:1 ratio of tall to dwarf peas.

INCOMPLETE DOMINANCE

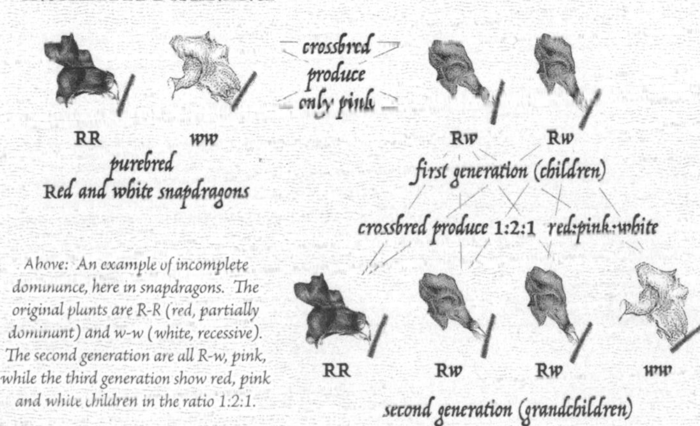

Above: An example of incomplete dominance, here in snapdragons. The original plants are R-R (red, partially dominant) and w-w (white, recessive). The second generation are all R-w, pink, while the third generation show red, pink and white children in the ratio 1:2:1.

Chromosomes

genes and DNA

Towards the end of the 19th century, scientists began turning their microscopes on cell nuclei, looking for the components responsible for the evolutionary mechanism, and coining the term *chromosome* for the stripy pillules they could see in the nucleus. Observations of cell division (*mitosis*), gamete production (*meiosis*) and fertilization showed that chromosomes behaved in an organised way and it was soon suggested that they might carry inherited information as strings of heredity particles. By the 1920s the black string inside chromosomes was revealed to be chains of base/sugar/phosphate nucleotides, deoxyribonucleic acid, or 'DNA'. Its extraordinary double-helix structure was finally discovered in 1953.

DNA is a 4-letter code for life, universal to all life on Earth, i.e., all organisms use it in exactly the same way. The number of chromosomes varies from species to species (*see opposite top*), but all animals carry two versions of each chromosome in every cell nucleus, one from Mom, one from Dad, and spaced out along every chromosome are special sections of DNA called *genes*.

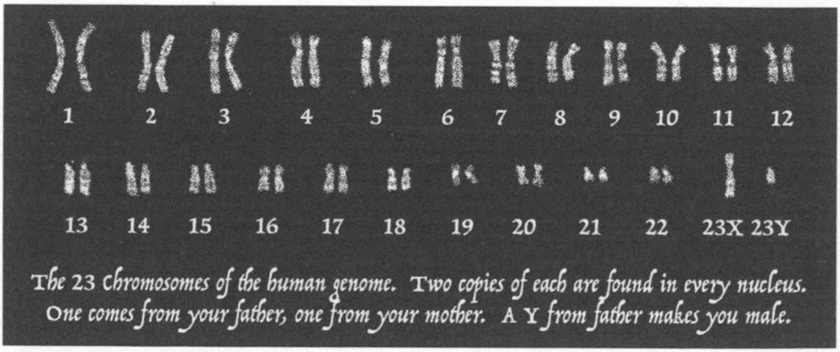

1 2 3 4 5 6 7 8 9 10 11 12

13 14 15 16 17 18 19 20 21 22 23X 23Y

The 23 chromosomes of the human genome. Two copies of each are found in every nucleus. One comes from your father, one from your mother. A Y from father makes you male.

ANIMALS

3 MOSQUITO 6
4 DROSOPHILA 8
6 HOUSE FLY 12
12 SALAMANDER 24
13 LEOPARD FROG 26
16 ALLIGATOR 32
20 SHREW 40
20 SQUIRREL 40
22 BAT 44
22 PORPOISE 44
23 HUMAN 46
27 GARDEN SNAIL 54
28 ELEPHANT 56
30 GOAT 60
32 ARMIDILLO 64
32 GUINEAU PIG 64
32 OPOSSUM 64
32 PORUCPINE 64
35 CAMEL 70
37 CHICKEN 74
39 DOG 78
41 TURKEY 82
66 KINGFISHER 132
104 KING CRAB 208

PLANTS

7 PETUNIA, X2, 14
7 PEA, X2, 14
7 LENTIL, X2, 14
7 RYE, X2, 14
7 EINKORN WHEAT, X2, 14
7 DURUM WHEAT, X4, 28
7 BREAD WHEAT, X6, 42
8 ALFALFA, X4, 32
9 LETTUCE, X2, 18
10 CORN, X2, 20
11 BEAN, X2, 22
11 MUNGBEAN, X2, 22
12 POTATO, X4, 48
12 TOMATO, X2, 24
12 RICE, X2, 24
12 PEPPER, X2, 24
14 APPLE, X2, 34
14, BRAMLEY APPLE, X3, 52
20 SOYBEAN, X2, 40
24 TOBACCO, X2, 48
41 LILY, X3, 82
630 FERN, X2, 1260

Above: Chromosome numbers in certain animals and plants. All animals are diploid, which is to say they carry two copies of each chromosome in every cell nucleus. For example, bats have paternal and maternal copies of 22 chromosomes, so 44 in each cell. Plants can be polyploid, having more than two copies, so triploid (three copies, normally infertile crossbreeds), tetraploid (four copies), or, even, hexaploid.

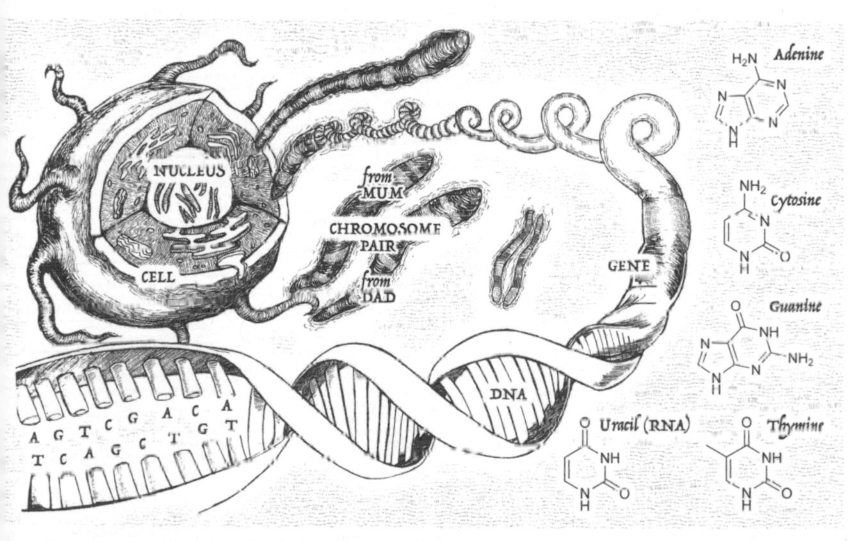

Above: Chromosomes live in the nucleus and are made of DNA. The double helix structure, like a twisted ladder, is comprised of pairs of just four bases, adenine (A) always bonding with thymine (T), and guanine (G) with cytosine (C). DNA thus stores information in a double-binary or quadrinary fashion, with each of the two strands being an identical copy of the other. A few thousand genes are spaced out along each chromosome, with long sections of repetitive non-coding DNA creating space between them.

The Book of Life

four letters, twenty words

The genome of a species is the entire DNA sequence found in its chromosomes. The human genome is like a cookbook as long as 1000 bibles, with 23 chapters (chromosomes), each chapter containing several thousand recipes (*genes*). Each recipe is for one protein, and is written using just 20 different words (*codons*), made of only four letters (*bases*). The recipes have advertisments (*introns*) in them, which have to be snipped out for the finished copy (*exons*).

When the human genome was mapped in 2000, scientists were surprised to find it contained pages and pages of gobbldygook *between* each gene. Some sections of these non-coding (or "junk") DNA sequences originate from distantly broken genes, while others are repetitive transcription errors (DNA can lose count when copying repetitive sequences like TATATATA). Other sections are dead retroviruses (viruses which use reverse transcriptase to copy their RNA into their host's DNA, so that they become part of their host's genome). Another group of genetic parasites, descended from retroviruses, are known as "jumping genes." These useless little sequences are found as introns in almost every gene and shout "copy me everywhere" to passing chemical equipment. Highly virulent little data bugs, today they make up about a quarter of our DNA. "Real" genes account for only 3%.

Non-coding DNA has its uses. It creates spaces between genes, aiding clear transcription, and preventing their breaking during *crossover* (*page 14*). Though not expressed as proteins, non-coding DNA can also finely regulate genetic expression, enhancing or suppressing the transcription of genes it adjoins.

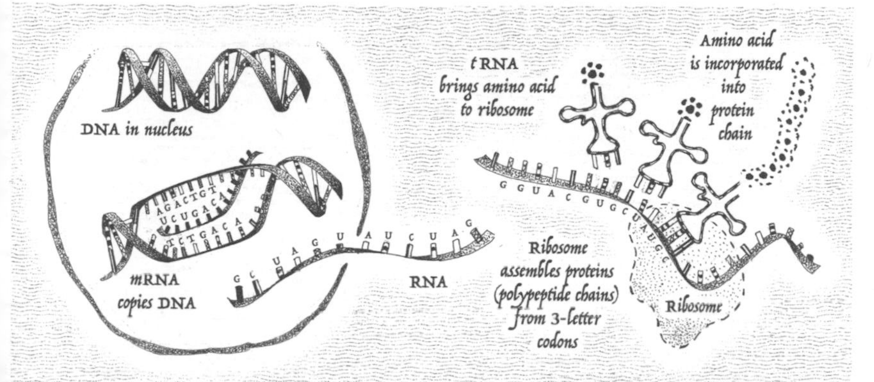

Above: DNA in the nucleus, comprising of adenine (A)-thymine (T) and guanine (G)-cytosine(C) bonds uncoils and is transcribed into a strand of messenger RNA, identical to DNA except for the replacement of thymine (T) by uracil (U). The transcription unit is read in three letter words, each of which code for an amino acid. These are then converted into a string of amino acids known as a polypeptide chain, or protein.

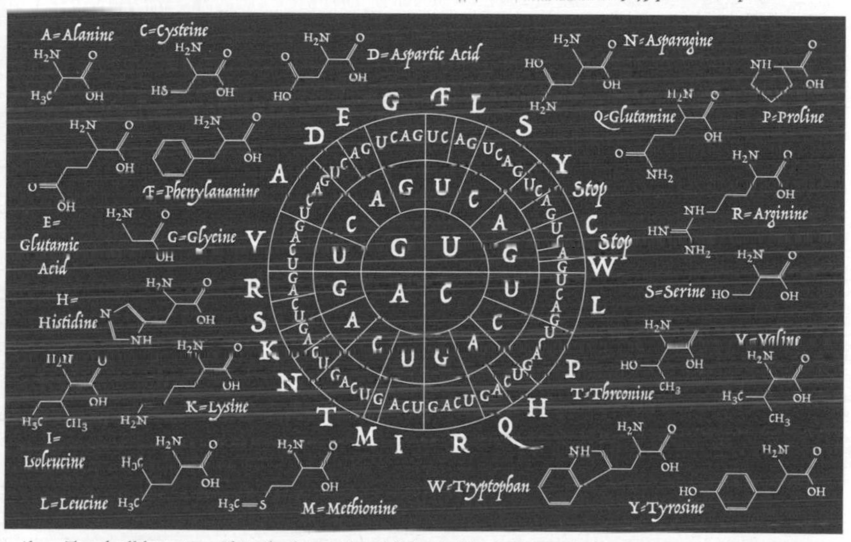

Above: The code of life is written with just four letters and words (codons) of three letters. Use the chart above, starting at the centre, to find out which amino acid is produced by any word. For example UAC leads to Y, Tyrosine. Most amino acids are coded for by more than one word, so that there are only 20 amino acids used by life. All life on Earth, whether tree, beetle, rabbit, human, or fungus, uses exactly the same code.

A World of Variation

how it all gets mixed up

Darwin's theory depended on a mechanism that could create small variations in offspring. The answer was hidden deep within the way *gametes* (sperm, eggs, or plant spores) are produced.

As Mendel had guessed, a complete copy of all the DNA needed to build an entire organism is held in every cell nucleus, organised into discrete chromosomes, with one *homologous* (similarly mapped but different) copy from each parent. During *meiosis* (*opposite*), genes are snipped out and swapped between the parental pairs, the reshuffling and recombining creating new chromosomes for the gametes, each with different trait potentials (*crossover shown below*).

This was one source of the variation Darwin had been looking for. Indeed, shuffling is all very well, but wild cards also play their part. During meiosis all sorts of things can go awry: copying errors, occasional deletions, duplications, sometimes inversions of sections of DNA, sometimes in genes, but more often in non-coding DNA. Often subtle (the tiniest change to a single letter or protein), they occasionally turn out to be advantageous, but can also be fatal.

There is a gene on chromosome 4 which simply contains the word "CAG," repeated again and again. Most people have it repeated from anywhere between 6 and 30 times, but if you have it repeated over 35 times you will slowly die of Wolf-Hirschhorn syndrome. Another example: A single-letter change in a 253-word gene on chromosome 20 will give you mad cow disease.

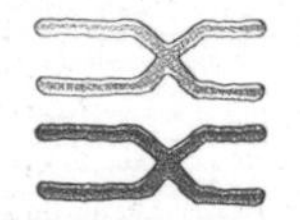
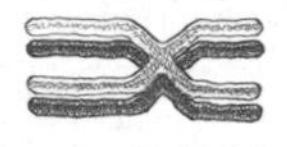

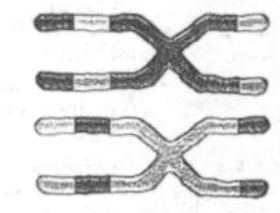

MITOSIS = CELL DIVISION

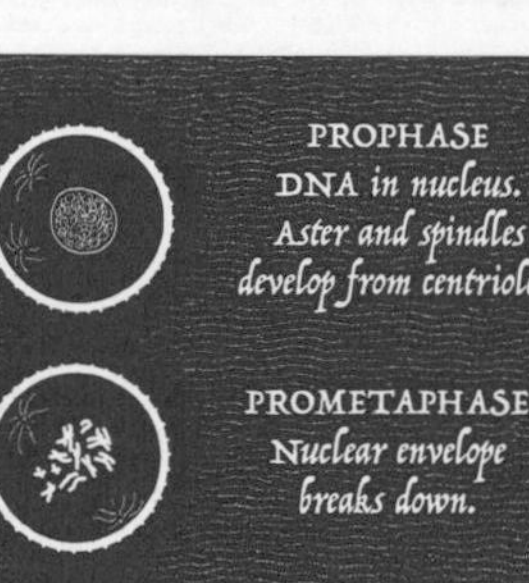

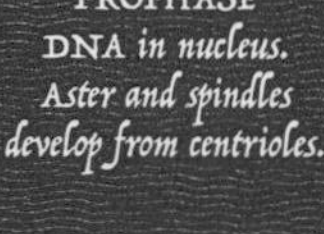

PROPHASE
DNA in nucleus. Aster and spindles develop from centrioles.

PROMETAPHASE
Nuclear envelope breaks down.

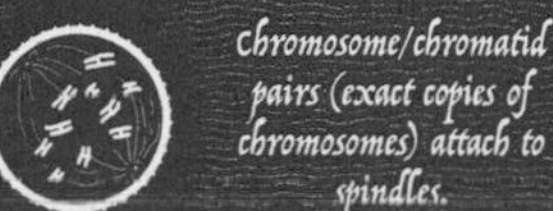

Chromosome/chromatid pairs (exact copies of chromosomes) attach to spindles.

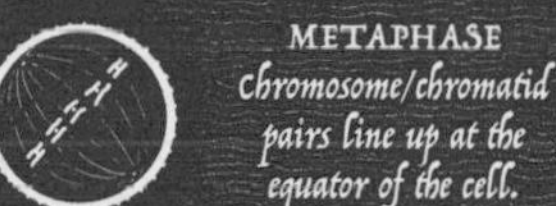

METAPHASE
Chromosome/chromatid pairs line up at the equator of the cell.

ANAPHASE
The spindles pull the chromosomes and chromatids apart.

In humans, 46 chromosomes are pulled into each half.

TELOPHASE
Chromosomes reach mitotic poles and cell starts to pinch.

CELL DIVISION
followed by

INTERPHASE: *chromatids created for next division.*

MEIOSIS = PRODUCTION OF SPERM OR EGG

PROPHASE I
Maternal and paternal copies of chromosomes pair up.

Each chromosome replicates, creating a chromatid.

CROSSOVER *occurs between maternal and paternal versions*

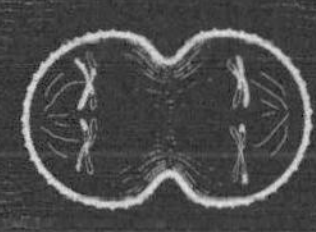

METAPHASE I
Shuffled chromosome/chromatid pairs are separated

TELOPHASE I
Division into two

PROPHASE II
Division
METAPHASE II

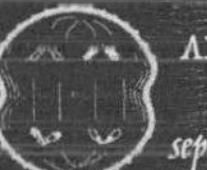

ANAPHASE II
Further separation of pairs

TELOPHASE II
Creation of four haploid cells, each with one copy of each chromosome

Nurturing Nature

Baldwinism, and the behavioral sieve

In 1896 James Mark Baldwin (1861-1934) advanced a theory that learned advantageous behaviors could eventually become instincts. He proposed that behavioral, cultural and even chemical factors could greatly shape a genome. His ideas were ahead of his time, as he had effectively predicted the modern study of memetics (*page 42*) and epigenetics (*page 18*).

Organisms possess genes that incline them to behave in certain ways (and in certain ways at certain times). When circumstances change, which they invariably do, individuals more inclined to acquire appropriate new behavior patterns will survive better, breed more, and amplify those inclinations (*see opposite*).

The debate between those who believe that habits are primarily dictated by DNA and those who believe that they are mostly learned has become ever more tangled. Today, many instincts are understood as needing to be triggered *via* appropriate nurture. For instance, a monkey reared in the wild learns the fear of snakes from its mother (perhaps she screamed on seeing them). A young monkey reared with no knowledge of snakes quickly learns to be scared of them if adults communicate fear in the presence of one.

But when an adult monkey trained to be scared of flowers (don't ask how) is introduced to a young monkey reared in captivity, he can scream at flowers all day but the young monkey will just look at him as if he is crazy. There is no innate triggerable fear of flowers. There is one for snakes. Many genetic instincts work like this, and lie dormant until switched on or accessed. Only certain patterns at the right time will cause them to come to life.

Above: A group of frogs, recently hatched from a small pool, attempt to cross a river. Different frogs have different inclinations, causing them to learn to either cross a log, jump lily pads, or head for the branches overhead.

Above: Natural selection takes its course. The slight differences in behavioral inclinations have favored the frogs that learned to climb trees. All the others fall victim to predators in the river. The tree frogs get to breed.

Above: The useful behavioral inclinations spread through the gene pool and become instincts. New behavioral inclinations and other examples of variation begin to appear and the process begins again.

Epigenetics

same genes, but different ones expressed

In 1942 Conrad Hal Waddington [1905-75] described a new science, *epigenetics*—a branch of biology that studies those causal interactions between genes and their products which bring the phenotype into being. The term today refers to heritable traits which can be passed down from parents as *expressions* of certain genes (using a process known as methylation, where small markers stick to the parental DNA). Despite there being a full copy of parental DNA in every nucleus, only a small proportion of genes are active, or switched "on" in any cell at any one time. The bloodstream is full of millions of different chemical messengers that target specific genes in specific cells. By noticing and remembering which genes that were "on," or "off," sperm and eggs can pass on certain acquired parental traits. A mechanism for Lamarckism.

Environmental conditions, diet and pollution, have all been shown to influence genomic changes in offspring. What you do today really does influence the genome of your great-great grandchild. The epigenetic process has been likened to chewing-gum stuck on the DNA switches. The gum can be either "on" or "off" at any time, inhibiting the switching of the gene. If both gene and gum are "on," the gene has to stay "on" until the gum is removed (by an external influence); and vice versa, and in combination.

Epigenetics reveals that emotions, fears, addictions and other triggers and hormonal surges (that course as a chemical cocktail through your arteries) can all be passed on to your children by this process. Additionally, you really can affect the *expression* of your DNA simply by thinking about things!

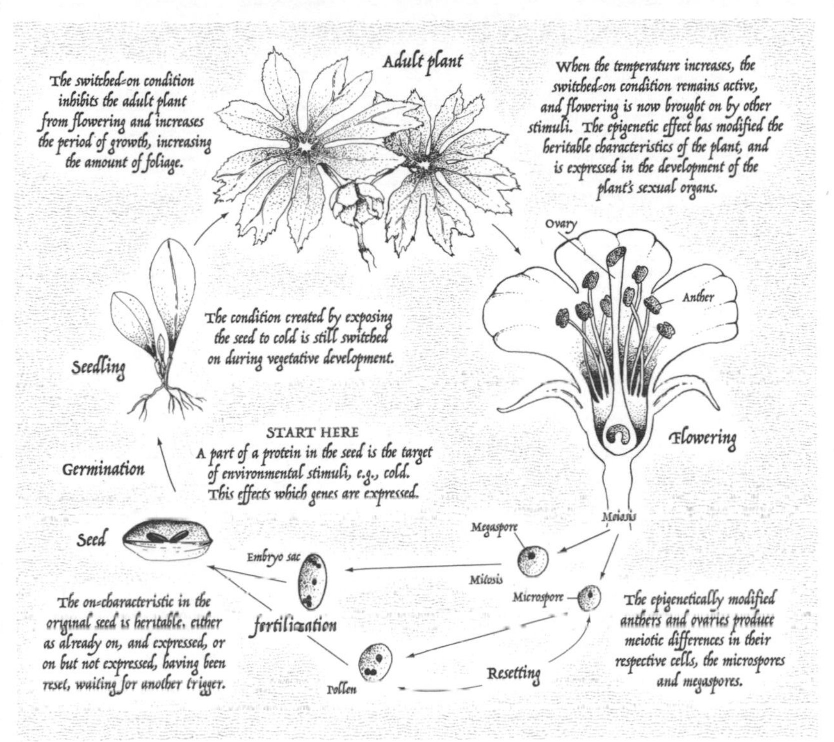

Above: Epigenetics is the study of gene expression. Genes may be unalterable during a lifetime, but their particular expression is affected by many factors and can be passed on to descendants. Genes can be switched on, or off, depending on the cocktail of proteins and hormones flooding the system, and their sensitivity to the cocktail. In the example above, the plant's chemical response to exposing the seed to cold can be passed on to its own offspring.

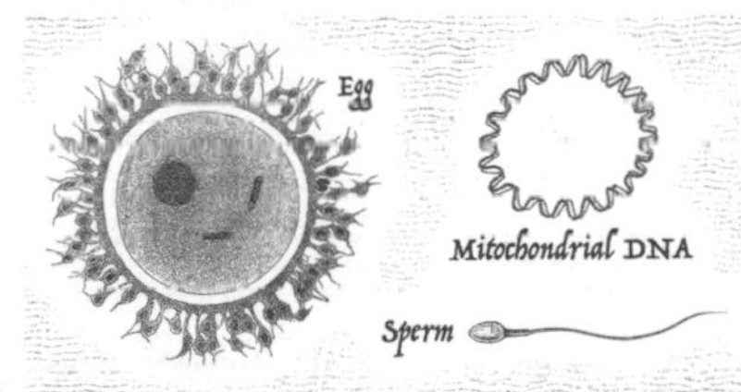

Left: Although a father delivers his DNA to his children, only the nuclear DNA makes it through. The DNA for your mitochondria (the bacteriological batteries in your cells), and for the chloroplasts in plants, as well as the chemical soup of intracellular fluid, the vacuoles, and much of the garden that surrounds and supports the nucleus are all inherited only from your mother. Thus mothers can pass on more epigenetic triggers.

The Red Queen

evolutionary arms race

All species are in constant competition with others for resources, and one result of this is that they all need to keep evolving just to maintain the status quo. In species that relate as predator and prey, sharper teeth or greater speed in a predator may result in thicker armor or faster legs in its prey. The concept was first described by Leigh Van Valen in 1976, and termed the *Red Queen* effect, after Lewis Carroll's *Through the Looking Glass*, when the Red Queen remarks to Alice, "it takes all the running you can do, to keep in the same place," It turns out that perpetual motion is a prerequisite of evolution. Because environmental conditions are always in flux, so too are the organisms that populate them (*see examples opposite*).

One example is the role of sex in fighting disease. Diseases break into cells, either eating them (fungi and bacteria) or taking over their genetic machinery (viruses). They get in by using protein keys, and successful break-ins can lead to the key spreading fast. Sex, as opposed to cloning, creates children who are different to one another, and have a variety of different locks to keep the parasites guessing. For example the flax plant has 27 versions of 5 different genes that help resist rust fungus, and different individuals have different combinations. Resistance genes that work become widespread, but then so do parasites that unlock them, and then new corresponding resistance genes, then new keys, and so on.

The pace of evolutionary change varies. *Saltation* emphasises the role of mutations in the sudden morphological changes that result in branching of the evolutionary tree, while *gradualism* emphazises natural selection and the subtle adaptation of species over time.

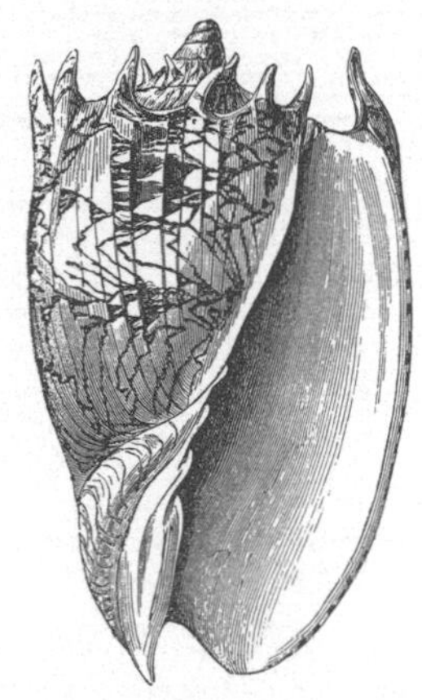

Above: Many predator/prey systems engage in arms races. For example, over millions of years many mollusks have evolved thick shells and spines to avoid being eaten by animals such as crabs and fish. These predators have, in turn, evolved powerful claws and jaws that compensate for the snails' thick shells and spines.

Left. In the arms race between plants and insects, plants that evolve a chemical that is repellent or harmful to insects will be favored by natural selection. But the spread of this gene puts pressure on the insect population, favoring insects that evolve the ability to overcome this defense. This, in turn, puts pressure on the plant population, and any plant that evolves a stronger chemical defense will be favored. This, in turn, puts more pressure on the insect population ... and so on. The levels of defense and counter-defense perpetually escalate, without either side ever "winning."

SPECIATION

don't play with those children, Rose

When two or more groups of a species are separated and evolve in different directions so far that they can no longer breed with one another, scientists talk of *speciation*, the creation of new species. Speciation is often precipitated by the isolation of and subsequent readaptation of a part of the population. In the case of mankind it seems likely that events created a selection pressure. Two short ancestral ape chromosomes fused together into a new chromosome 2 in a single individual and in an isolated group (*below*). Chimpanzees today still have the original 24 chromosomes.

Darwin stuggled to explain how new species evolved from a single ancestral species when no physical isolation or barrier was involved (*allopatric* speciation). In fact, differing behavior and inclinations in subpopulations can be all that is required for them to genetically isolate themselves. A useful model is that of cichlid fish. In a population frequenting two favorable habitats separated by a barren area, fish inclined to remain within the favorable habitats stand a better chance of survival. Over time natural selection takes the two populations in different directions until they become sub-species and then entirely separate species, unable to interbreed.

Opposite and above: Speciation caused by isolation. Opposite: The ape ancestor of humans becomes separated by a huge rift. A single mutation in the isolated group fuses two chromosomes together (6 millions years ago) resulting in the creation of a new species. Above: Chimapanzee ancestors play together across a young Congo river. The Congo widens, separating two groups which evolve into modern tool-wielding common chimpanzees and the sex-obsessed Bonobo chimpanzees.

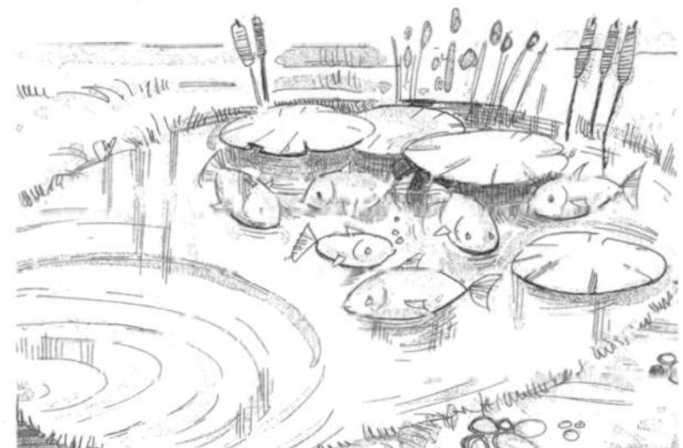

Above: Speciation in a pond. Fish evolve that prefer living under lily leaves. A new area of lilies appears across the pond. The fish show no interest in crossing the barren area between the two areas, so each evolves separately.

Above: Speciation in a mixed population. A varied species, with large, small, light and dark fish, begin to develop sexual preferences for their own size and colour. Eventually two distinct species evolve.

The Migration of Genes
out of Africa

How do the gene pools of a species behave over wide areas? Changes can often begin in areas isolated through physical or behavioral causes, so that localised subpopulations become hot-spots of new genetic information. Genes can also drift about within species, as individuals explore, travel and fall in love.

The evolutionary history of humanity shows combinations of both hot-spotting and ubiquitous drift in the fossil record, smoothly progressing in some periods, while in others it is more staggered. Hot-spotting has resulted in the many races of humans, while ubiquitous drift within those races has resulted in the relatively uniform genetic characteristics that can be seen.

The study of genes has enabled various maps of humankind to be plotted. Circular mitochondrial DNA, for example, is only inherited from mothers, so avoids the shuffling of meiosis, and remains virtually unchanged down the generations. Studies of mitochondrial DNA have revealed that 99% of Europeans are descended from just seven women (known as *clan mothers*) living at different sites in Europe at different times during the last Ice Age. Globally, all humans are now known to be descended from a single ancestress living in Africa about 200,000 years ago.

Similar studies of the Y-chromosome, which is only passed from father to son (also virtually unchanged), have revealed that 99% of Europeans are descended from just five men (*clan fathers*), living during the last Ice Age. All humans are now known to descend from just one man, living in Africa about 70,000 years ago.

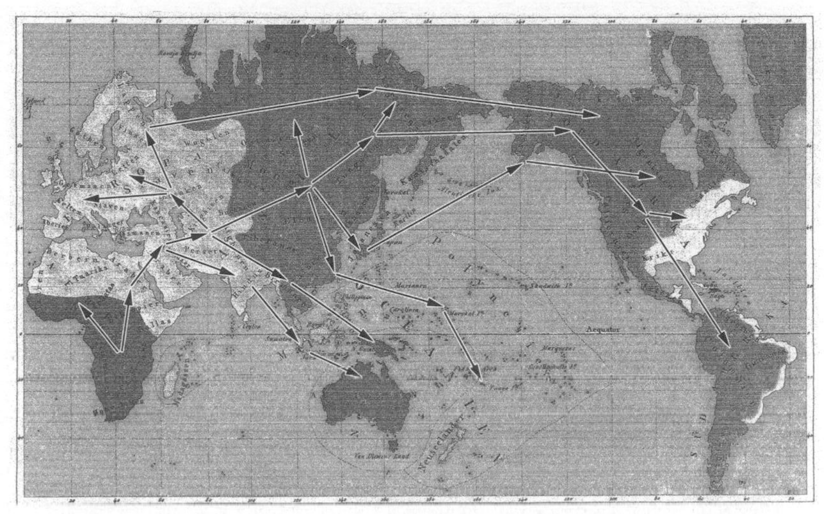

Above: Studies of mitochondrial DNA and the Y-chromosome have revealed the path by which humankind left Africa, by a northeasterly route, roughly 65,000 years ago. The ancestors split up, some heading south, others heading east, then northwest. Europe was populated around 40,000 years ago, and the Americas 25,000 years ago.

Above: Studies have revealed that we are all descended from a handful of ancestors. For instance, the Y-chromosome of Native American peoples has revealed a single Native American man, from whom 85% of all Native Americans in South America, and half those in North America, are directly descended.

Initiation and Cooperation

right from the start

The origin of the first strand of DNA or RNA on Earth remains a mystery. It may have arrived from elsewhere, but even so, somewhere in the universe, nucleic acids must have been conjured from a primal ooze, perhaps in hydrogel during lightning strikes, or in deep hot fissures, underground or underwater. Lipid bilayers, the basic structure of cell walls, spontaneously form from phospholipids, and one strand of nucleic acid seems to have had the correct coding for manufacturing these. Thus was born the first organism (*below*), which multiplied until its clones and variants began teaming up with or competing against one another. Symbiotic relationships (*see page 34*) between single-celled organisms like bacteria and archaea (*the two kinds of prokaryotes, see page 50*) became formalised when colonies of cells began sharing their DNA-strands in a nucleus, giving rise to more complex, multicellular organisms (*eukaryotes, pages 52-57*).

The story of life is, therefore, one of cooperation as much as competition. Importantly, cells are specialised for different tasks, much like human beings (indeed human culture may be seen as uniquely resulting from specialization, or the division of labor). By trying new kinds of cells, forms, partnerships and energy sources, DNA hosts managed to leave their watery origin and survive in a wide variety of habitats, always carrying the currently successful versions of the DNA code around with them.

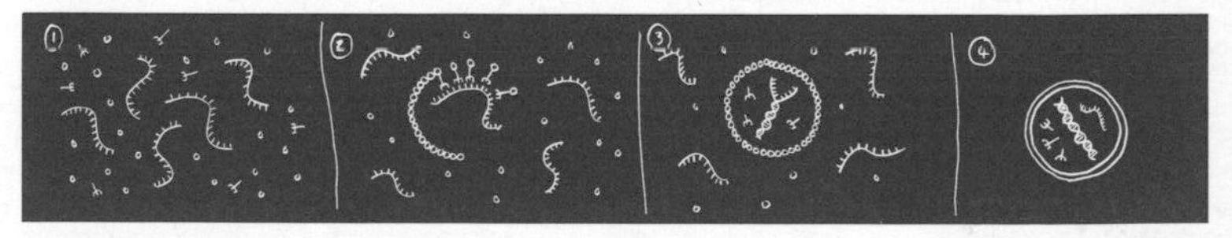

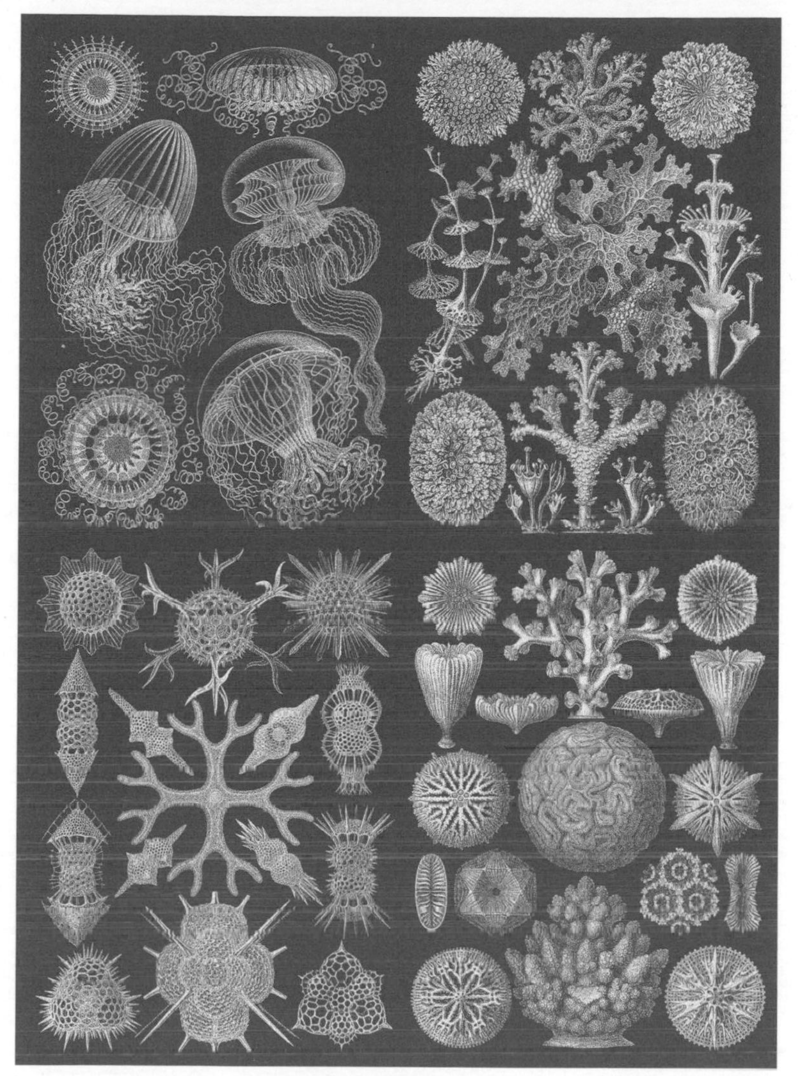

Parasitism and Symbiosis
the human question

Many organisms develop relationships with others, either *parasitic* or *symbiotic*. In parasitic relationships only one species benefits from the arrangement, while symbiotic relationships suit both or all parties. Lice, fleas and worms, for example, afford no advantage to their hosts, but gain resources themselves, while the bacteria in our stomach help us digest, and get fed in the process. Although some deadly bacteria and viruses (*below*) seem to have a lose-lose dynamic with their hosts, they aim to infect new victims before both die.

Some symbiotic relationships are so involved that they result in composite organisms. An example of an animal-animal composite is the Portuguese man-of-war, which looks like a jellyfish. The components of its body are actually different species of organisms working in a cooperative colony. An example of a plant-plant composite is a lichen, which is part alga and part fungus. There are also animal-plant composites, such as upside-down medusas (*top left, page 27)*, which are jellyfish that contain colonies of algae.

Mankind may be increasingly likened to a parasite with respect to most life on Earth. Meanwhile, we are in a symbiotic relationship with daffodils, apple trees, dogs, cows, chickens, grasses and a few other species, all of which thrive at the expense of the many.

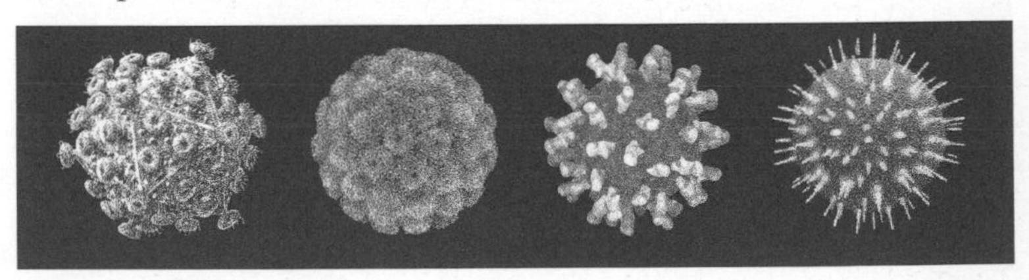

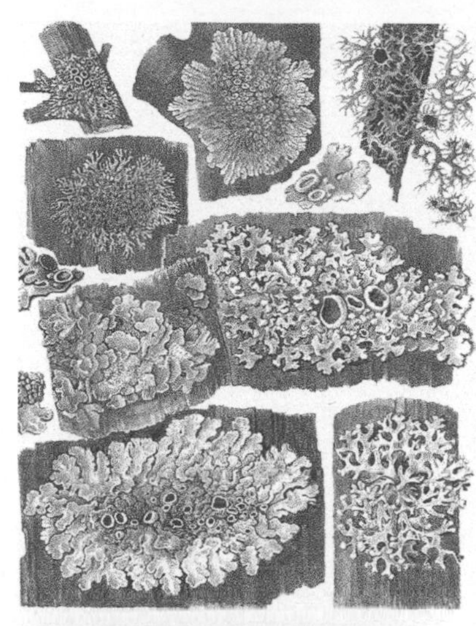

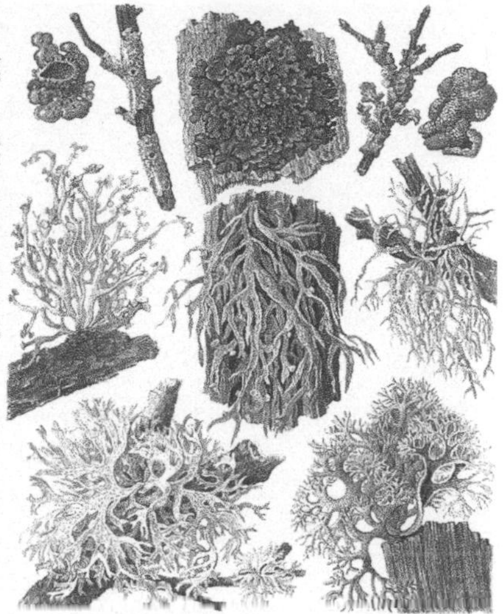

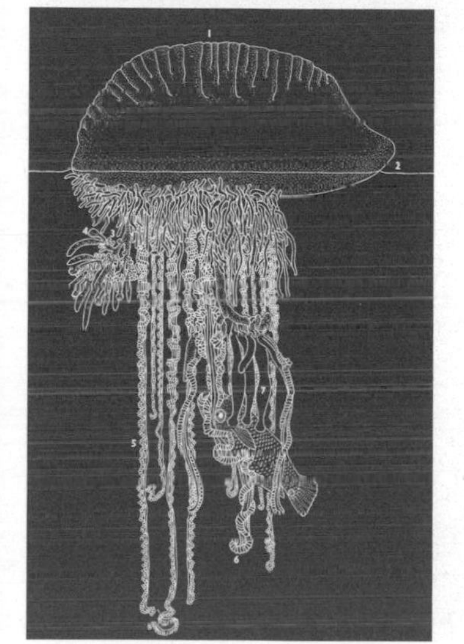

Above, and top right page 27: Various species of lichen. Lichens are composite organisms consisting of a symbiosis between fungi and photosynthetic partners which derive food for the lichen from sunlight.

Left: A Portuguese man-of-war. This is a siphonophore, a colony of specialised polyps and medusoids.

Below: The household flea. An example of a parasite, bringing no gain to its host, simply feeding off it.

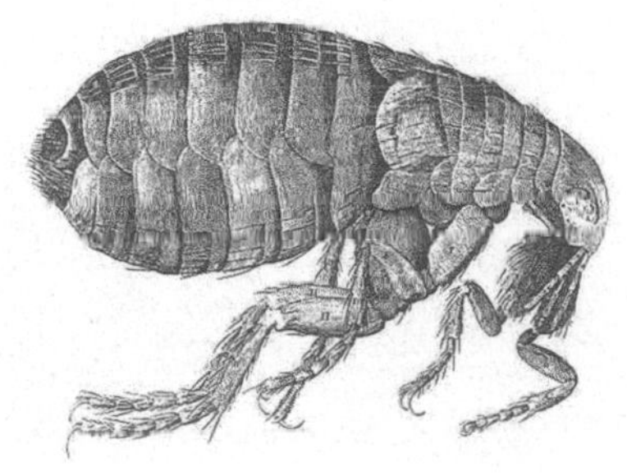

Kin Kindness

all for one and one for all

Many creatures perform seemingly selfless acts which protect the genome of their close kin, or even their entire tribe. Altruism operates as a mechanism in evolution because adaptations can also occur at the group level. Although personal selfishness tends to beat altruism within a group, altruistic groups generally beat selfish groups. Altruistic behavior ensures the survival of group characteristics, both genetic and memetic.

In social insects (ants, wasps, bees and termites), sterile workers devote their entire lives to the queen, with no chance of breeding themselves. Vampire bats (*opposite top*) share blood with hungry neighbors on returning to their caves, but keep a tally, and expect a return of the favor. Altruism has behavioral mottos: "family first," "help your friends and they'll help you," "safety in numbers," "care for the sick and elderly." A vervet monkey will make an alarm call to warn chums of an unnoticed predator, despite making itself a target.

Extending their sense of kin, dogs occasionally adopt orphaned cats, squirrels, or ducks. Dolphins occasionally assist sick, injured, or struggling animals, their extended evolutionary kin.

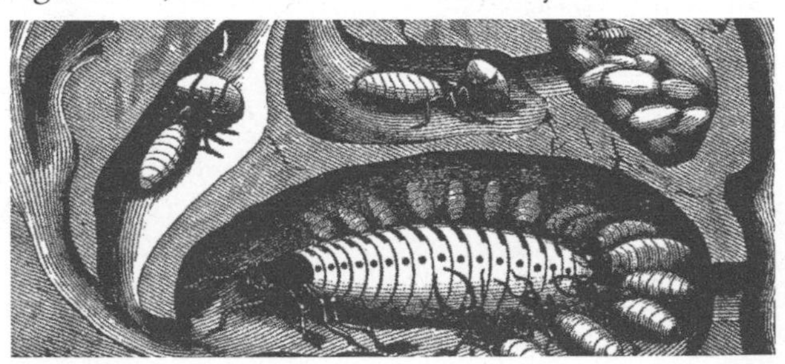

Above: Vampire bats exhibit one of the most universal kinds of altruism, reciprocity. Tending to roost in the same place every night, they get to know each other and will regurgitate blood for a hungry neighbor if the favor is likely to be reciprocated. Vervet monkeys, likewise, will come to the aid of others who have helped them in the past.
Below: Walruses will adopt and rear young calves from different families whose parents have died.

Sexual Selection

the beautiful things creatures find attractive

Sexual reproduction introduces variation beautifully, which is why most complex species use it. Even non-sexual species occasionally employ some form of sexual reproduction to avoid stagnation. Species that reproduce asexually (without meiosis or mitosis) or by parthenogenesis (without male fertilization) do not produce the genetic variation necessary for effective natural selection.

Males in some animal species hope to mate with as many females as they can, ensuring the wide dispersal of their DNA through their billions of sperm. Meanwhile, a female who has to devote significant energies to the survival of her offspring has fewer opportunites to pass on her DNA. Because she can only have a small number of offspring, it is generally in her interest to look for the best DNA available. As a result she is often choosier than a male, and can require proof of quality, or seduction. This can lead to amazing results.

The classic example is the peacock's tail (*opposite*). It is a totally cumbersome disadvantage to peacocks in every respect apart from its sexiness to peahens. The fittest peacock with the tidiest and most mesmerising display gets to pass on his genes.

Female guppy fish find colourful males irresistible, especially those with prominent collars. So sexual selection gradually produces populations of colourful fish until the time that predators move in and easily target them. Natural selection leaves the duller ones.

In stag beetles (*opposite*) the outsized and largely useless mandibles are the product of sexual selection by females. They are used by the males to fight over the females, but the males select themselves, the females mating with whomever wins.

Convergent Evolution

inevitable solutions

Variation generally produces evolutionary divergence, as species adapt to fill as many econiches as they can. However, it turns out there are often only a limited number of good design solutions when it comes to solving problems such as how best to fly (*below*) or see (*opposite top*), and that is why many species share similar features. When not due to a directly shared ancestry (i.e., coded for by the same DNA) this is described as *convergent* evolution, as analogous features have evolved completely independently.

Convergent evolution can result in both collective and singular analogous traits. In marsupial and placental mammals, corresponding species that have adapted to fill similar econiches on different continents generally resemble one another despite being genetically distant. An example of a singular analogous trait is the eye of the cephalopod and the eye of the vertebrate. Their eye-coding DNA is different, yet their eyes are virtually identical in both structure and function. Groping around for improvements, evolution again and again finds the same solutions that work best.

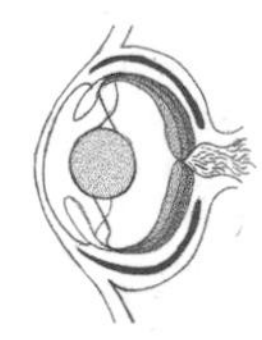

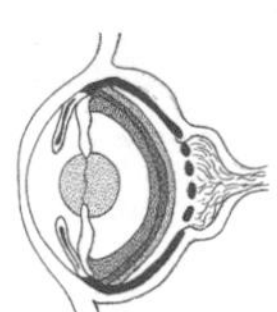

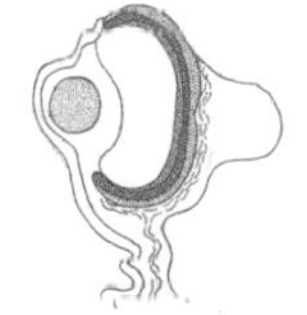

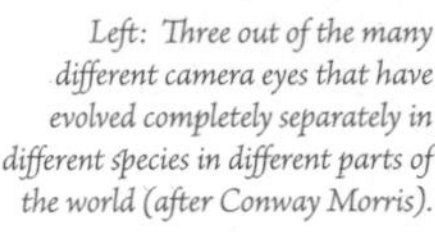

Left: Three out of the many different camera eyes that have evolved completely separately in different species in different parts of the world (after Conway Morris).

At the top, the human eye focuses by changing the shape of its lens. In the centre, the eye of an octopus (a cephalopod mollusk) focuses by moving its lens forward and back. At the bottom, the eye of a marine annelid, a relative of the earthworm. Lens eyes have also evolved independently in the brainless cubozoan jellyfish, dinopis spiders and heteropod snails, to name but a few examples.

Optic nerve
Retina
Pigmented layer
Nuclear layer

This page, and opposite: Examples of convergent evolution. In the matrix of all possible solutions to problems, only some work, and only some of these work well. Whether it is the discovery of a chemical which responds to light, an efficient method of propulsion, or an outer form maximised for lack of drag through water, the best solutions are limited in number, and constrain the outcomes. The idea is not dissimilar to Plato's ideal forms, perfect solutions, shadows of which appear on Earth.

Above: Marsupial mammals have independently evolved similar forms to non-marsupials. There are marsupial deer (kangaroo), squirrels (koala), rabbits (bandicoots), rats and mice.

Above: Sharks and crocodiles are perfected forms which have changed only slightly in the last 100 million years. Niche species in Lake Tanganyika have evolved to ressemble ocean counterparts.

DEATH

and other helpful illnesses

Death is something everyone can be sure of. However, cells can theoretically reproduce infinitely (they have been kept alive for decades in laboratories), keeping the body young and strong, so why are they programmed to stop? Each cell can only replace itself a certain number of times. The protective tip of each chromosome (*the telomere, opposite top*) shortens on each copying, like a fuse, and when it is gone, decay and death follow. As we get older our DNA accumulates errors, and death stops these being passed on. Sex and death drive evolution. Thus, while women safely create all their ova before they are born, men's sperm become ever more error-ridden as they age. Even long-lived tortoises (150 years) and crocodiles (100 years) occasionally need to be refreshed from their gene pools.

Some illnesses also confer advantage. A famous example is sickle-cell disease, which protects against malaria (*lower opposite*).

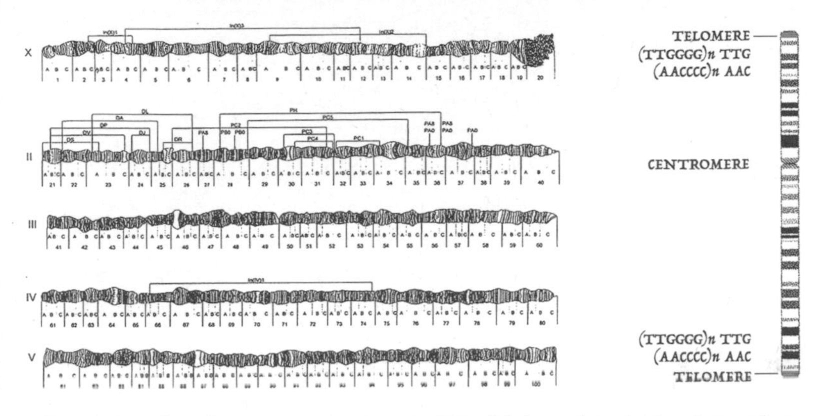

Above: At the tip of every chromosome are regions of repetitive DNA called telomeres that protect the ends. In vertebrates, fungi and even some slime molds, the repeated motif is TTAGGG. In insects it is TTAGG. The number of repeats varies (around 1000 in humans). Every time a cell divides, the copying mechanism misses some repeats, and shortens the chromosome. When is it all gone, the cell cannot reproduce, only die. Death is part of the program.

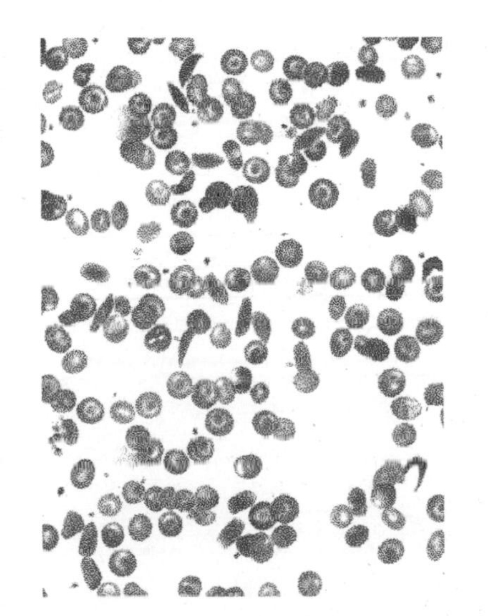

Left: Sickle-cell disease is a genetically inherited condition which reduces life expectancy because the deformed sickle-shaped red blood cells it produces inhibit gaseous exchange in the lungs. The disease reduces the life expectancy of those who have it, so why it has persisted? The answer is that it confers a resistance to malaria, transmitted by mosquitos (above). So in places where people are more likely to die from malaria than sickle-cell, populations tend to have unusually high instances of the disease.

Mimicry and Camouflage

the advantages they confer

Many animals use nature's own visual language: scaring away predators by looking more dangerous than they really are, or disappearing into the background, vanishing from hungry eyes.

Camouflage is a form of mimicry, where animals evolve to mimic their surroundings to improve their prospects of survival as predator or prey. In animals its effectiveness can rely on both the correct appearance and appropriate behavior.

Other forms of mimicry involve species mimicking each other. *Batesian* mimicry is where harmless species mimic harmful ones. Wasps, for example, are harmful and have aposematic (warning) stripes in black and yellow. A number of moths, beetles and hoverflies mimic them, resulting in birds steering clear of them for fear of being stung. In *Müllerian* mimicry, species mimic one another for mutual benefit. It is seen in similarly marked tropical butterflies, which are all distasteful to birds. In *Mertensian/Emsleyan* mimicry, deadly prey mimic a less dangerous species. This is because they are so poisonous that predators always die from their bites, never getting a chance to learn to avoid them. Certain deadly coral snakes mimic other snakes that are less harmful (*below*).

Mimicry is occasionally seen in plants too. Some tropical vines have fake butterfly eggs on their leaves, so that female butterflies lay their real eggs elsewhere.

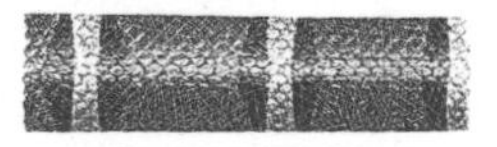

King Snake = harmless

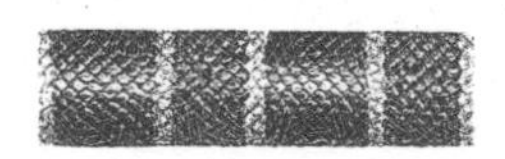

Coral Snake = deadly

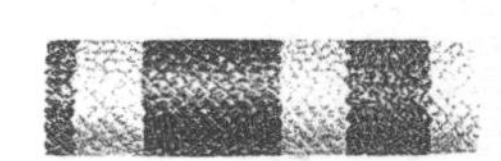

Arizona Coral Snake = poisonous

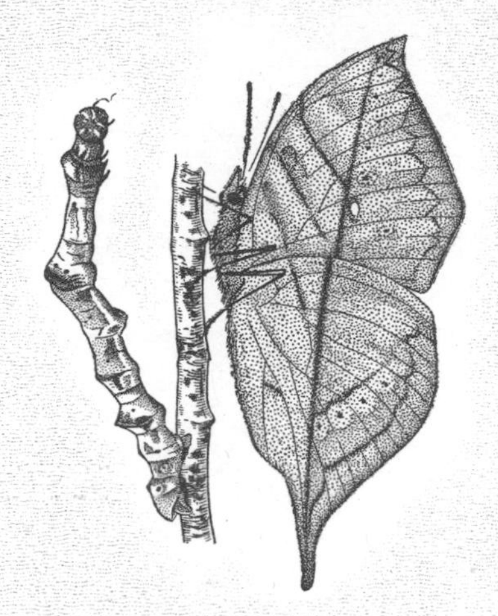

Above: Batesian and Müllerian mimicry. The harmless hoverfly finds it useful to have the warning markings of a stinging tree wasp—an example of Batesian mimicry. The potter wasp displays warning stripes, not dissimilar to those of tree wasps, an example of Mullerian mimicry.

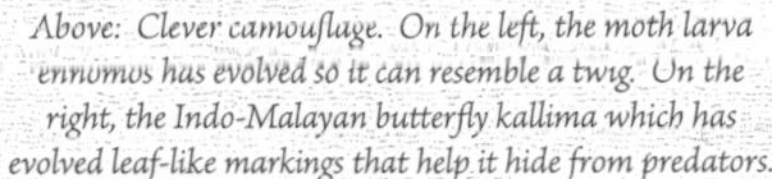

Above: Clever camouflage. On the left, the moth larva ennomos has evolved so it can resemble a twig. On the right, the Indo-Malayan butterfly kallima which has evolved leaf-like markings that help it hide from predators.

Above, and camougflaged left: The sargassum weed-fish has evolved extraordinary markings, protrusions and appendages, all of which make it extremely difficult for its predators to spot when it hides amongst the floating sargassum weed in the open zone of the Sargasso Sea.

It's Impossible
how does nature do it?

Opponents of evolutionary theory sometimes point to features that seem to deny the possibility of their having been through an evolutionary progression. Examples are the carnivorous leaf of the Venus flytrap and the mammalian eye. The argument is that they either work or they don't, there can be no halfway development.

But imagine evolution in reverse. The mammalian eye is a lens held in front of a retina by a sac of fluid. Imagine that sac of fluid getting progressively smaller, so the lens eventually sits on the retina. The lens and retina then fuse and the number of photosensitive cells reduces down to one. The result is a simple eye, common to many different organisms. It can still detect a movement around it, for instance an advancing predator blocking out the light.

In the case of the flytrap leaf (*lower opposite*), imagine the trigger mechanism removed. It may still trap the occasional tired old fly, but what if the leaf now exuded a glue to hold its prey while its lobes very slowly closed? Going back further, the finger-like appendages reduce and the lobe hinge disappears, so that the leaf can only fold itself by rolling. Earlier still, the surface area of the gluey substance may be optimised by forming droplets on hair-like protrusions. The result is the leaf of a sundew (*below*).

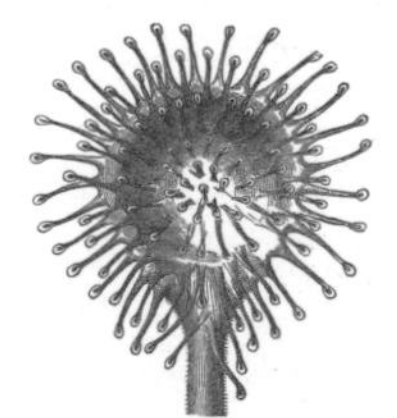
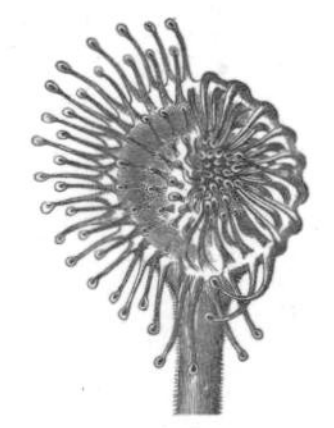

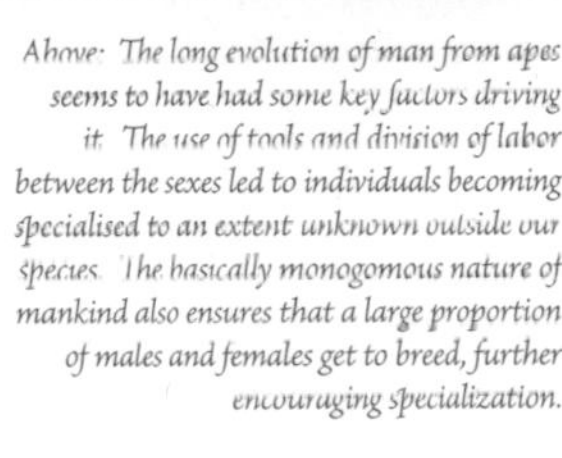

Above: The long evolution of man from apes seems to have had some key factors driving it. The use of tools and division of labor between the sexes led to individuals becoming specialised to an extent unknown outside our species. The basically monogomous nature of mankind also ensures that a large proportion of males and females get to breed, further encouraging specialization.

Left: The Venus fly trap. An example of the kind of organism used to oppose evolutionary theory. How could it have come about? Answer: Small refinements and improvements over a very large number of generations, with occasional larger mutations, each conferring advantage to its descendants, while the failures remain hidden by history.

MEMES

self-replicating thoughts and cultural viruses

Using biological evolution to help better understand cultural evolution, Richard Dawkins, in 1976, invented the concept of the *meme* as the cultural equivalent to the gene, defining it as unit of cultural information existing within a *meme pool.*

Memes manifest from thoughts or discoveries, and either survive in the meme pool or die off, depending on their value as perceived by individuals. Memes may also be represented in cultural patterns. A selection, perhaps, for a preference for a particular way of dressing, dining, or dancing, sends out cultural information arrays, which offer differing benefits to each individual that comes into contact with it. As this information feeds back between individuals and their groups, the effects spread, and subcultures form.

Memetics, the study of memes, describes how fashions come and go in what are often compared to viral or contagious effects. Other examples of memetic patterns include words, songs, phrases, beliefs, trends, habits and so on. Meme-gene coevolution, or *Duel-Inheritance Theory*, explains the ecology responsible for some of our ways of doing things and their genetic influences. Lactose tolerance, the Western acquisition of an adaptation involving increased tolerance to cow's milk, is just one example.

Other theories suggest stranger forces at play. Rupert Sheldrake, in studies from 1999-2005, showed that many people can sense when they are being stared at. His theory of morphic resonance proposes that similarly shaped ideas sometimes travel instantaneously between similarly shaped homes. Memes could be traveling between minds as quantum synchronicities in a holographic universe.

Memes are everywhere, affecting many of your actions.

Like genes, they can survive intact for centuries.

Some memes express themselves in trendy dress or speech.

Others are injected into society by advertisers and politicians.

Memes can take over what we notice, and what happens.

Folk wisdom is a fine kind of meme.

Accelerated Evolution
genetic engineering and self-evolved code

Humans are Gaia's best bet for colonising the galaxy so far. Will we continue becoming ever more intelligent? Or could memetic agents select against brains (brawn can turn nasty on brains). Would we survive an asteroid collision? Genetic engineering (*lower opposite*) may accelerate evolution in the not-too-distant future, creating enhanced lifespans, features and even subspecies. If we turn out to be not such a good bet after all, Gaia might try another trick. The eventual coloniser of our galaxy could evolve from some other source than apes (*dolphins, for example, opposite top*).

Evolutionary theory is used very effectively in the computer sciences. Dynamic programs spawn randomly varied children that are then selected for target behaviors. In this way robots effortlessly learn to walk in style, fly gracefully, or crawl and wriggle fast (*below*), using evolved algorithms which no human has programmed.

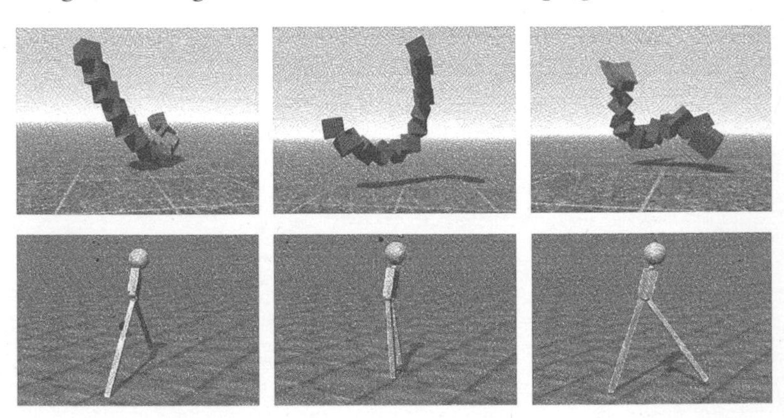

Above: Transgenic rice, engineered to address the serious iron and vitamin A deficiencies in developing countries.

Extraterrestrial Life

likely or not

There are approximately 100 trillion billion stars in the visible universe, so there may be plenty of planets with the right conditions for biological life. On Earth, now 4.6 billion years old, life started early, just 0.6 billion years after its formation, evolving either independently, or from bacteria trapped in cometary ice or space dust (a theory known as *panspermia*). These processes could easily have happened (or be happening) elsewhere.

DNA is not the only way to store large amounts of biological information, although it is one of the most efficient. Life might exist in sulfurous or silicate parallels to the versions we know. Other types of nucleic acid structures could have conjured themselves into cells and begun evolving elsewhere in the universe.

Despite internal differences, however, life on other planets is likely to outwardly exibit variations on familar themes, as genomic tribes find the best ways to stand, eat, collect sunlight, see, fly, swim and run. Similar rules of economy in form and function are likely to result in clades of organisms oddly familiar to us, as convergent evolution operates on a grand scale, and homologous organisms fill equivalent econiches to those on Earth. Gravitational variation might make legs shorter and thicker, or longer and thinner, but legs are still likely, and working legs, like eyes, have "best" designs too.

Above: An unlikely alien scene. Few of the creatures can walk properly, or see, or feed in the direction they are moving. Many have useless limbs or appendages. Natural selection is unlikely to have produced such clumsy life-forms.

Above: A slightly more likely alien scence. Animals traveling in liquids are fish-like. Plant-grazers resemble the horses and squirrels of Earth. Eyes look forward, dorsoventrality is the prevalent symmetry as on Earth.

The Evolving Biocosm

and the cosmological anthropic principle

The more we study the universe the stranger one single fact becomes. It turns out that not just the Earth, but our entire universe is nearly ideally suited to biological life. The physical constants that underlie the structure of space and matter are incredibly finely tuned to maximise the likelihood of fish, trees and things like us (*see examples opposite*). If any of the constants were different, often by a small degree, no life could exist in the universe. The conundrum, called the *Cosmological Anthropic Principle*, is one of the oddest products of modern big bang cosmology.

There are really only three answers to the problem. The first states that the tuning is a meaningless coincidence, the second holds that this must be just one of millions of universes (the one where things work out for life), and the third asserts that there must be a reason.

In 2003 James Gardner advanced a novel reason. He suggested that the beautiful fine-tuning we observe (and spring from) could be a kind of structural gene, passed on to a child universe from a parent or parents. Consciousness, life and its host forms could keep evolving to the point where an entire universe becomes a superorganism. This could then design or pass on a tuned set of constants to newly big-bang-born biocosms, guaranteeing them the best chance in life, as all organisms do for their children.

Quite what Darwin would have made of such a concept we can only guess at. But he might have been quietly pleased to see his theory remaining fit for purpose, adapting to the changing environment, and extending its reach to the possible variable, heritable and selectable nature of entire universes.

Fine Tuning in our Universe

Some life-friendly settings dating back to the Big Bang

1. Gravity is 10^{36} times weaker than the electromagnetic forces. If gravity was even slightly stronger, then stars, planets, and galaxies would be tiny and very short-lived. Life would be impossible.

2. If the strong nuclear force (which holds atomic nuclei together) were 0.1% weaker, then nothing beyond hydrogen would have formed in the universe. If it were 0.1% stronger, proton pairs would have taken over after the big bang, using up all the hydrogen in the universe.

3. If the expansion rate of the universe (which depends on many other factors) was slightly slower than it is, then the universe would have quickly recollapsed. If it was slightly faster, galaxies and stars could not have condensed out.

4. If there were greater irregularites or ripples in the early universe, then the universe today would be a much denser and more violent place. If the ripples were smaller, galaxies and stars would be flimsy, or not appear at all.

APPENDIX I - PROKARYOTES

The image opposite shows the family tree of life on earth. There are essentially two fundamental domains of life, firstly the subject of this page, single-celled microorganisms with no nucleus, called PROKARYOTES, which come in two important kinds, BACTERIA & ARCHAEA, and secondly those with multiple cells containing nucleii and bacteria (often mitochondria), which are collectively known as EUKARYOTES.

Although physically tiny, varieties of prokaryotes massively outnumber eukaryotes, as evolution has provided many more ecological niches for them to fill. Most of these free-living microorganisms or microbes cohabit environments alongside other organisms (or inside them) and have a fast life cycle (often dividing every 20 minutes), enabling them to diversify rapidly. Others (particularly archaea) survive in the more extreme habitats that were once ubiquitous on earth - some are known to have survived in salt crystals for around 250 million years. The structural simplicity of bacteria and archaea means that they are neither plant nor animal. Having no nucleus, their DNA is loose within their cell wall, and they can swap it freely with others, meaning that most bacteria can, in fact, be regarded as cells in a single global superorganism.

Bacteria are incredibly diverse. Some create spores or filaments, others glow in the dark, one or two turn milk to yogurt. Their classification is not complete, but here are some of the dozen or so phyla (kingdoms): Aquificae, Xenobacteria, Cyanobacteria, Proteobacteria (1650 different species), Firmicutes (2500 species), Spirochetes, Bacteroids, Flavobacteria, Fusobacteria, Thermomicrobia, Chlorobia, Sphingobacteria ...

Even more simple than prokaryotes are VIRUSES and PRIONS, which are regarded as non-living, despite their status as organic entities. Viruses are bundles of nucleic acid within a shell and cannot grow or reproduce outside of host cells. Most parasitise the cells of eukaryotic organisms. RETROVIRUSES transfer their DNA into the chromsomes of their hosts, others viruses invade bacteria and are known as BACTERIOPHAGES (bacteria-eaters).

Prions lack nucleic acid and shells. Being little more than particles of protein, they duplicate themselves either inside or outside the cells of the host organism.

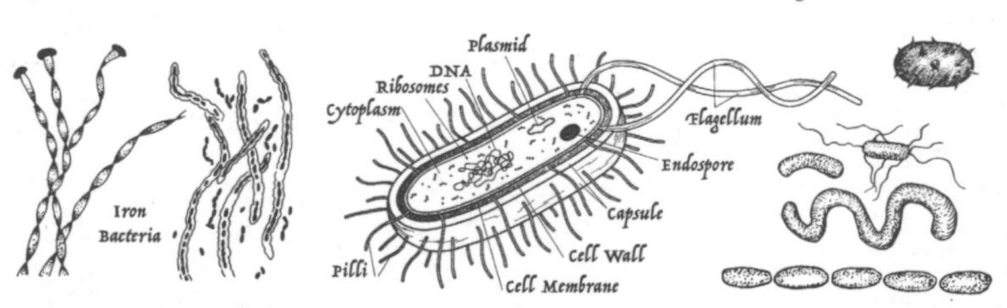

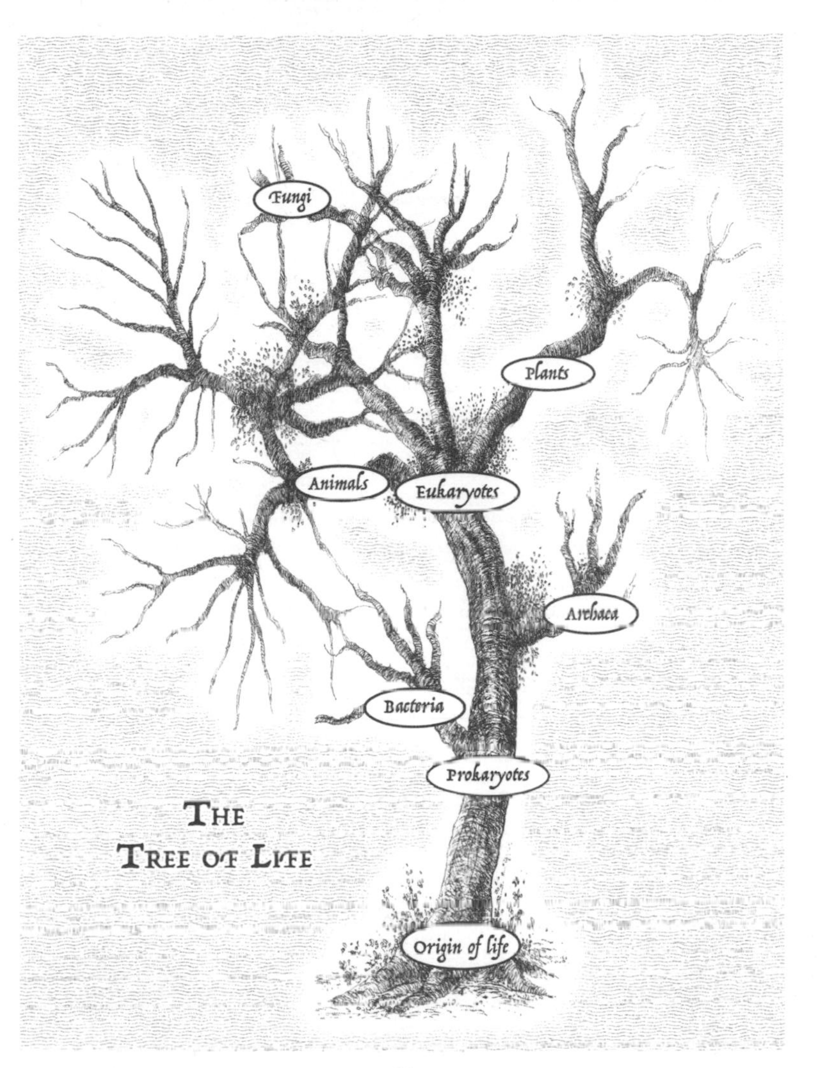
Fungi
Plants
Animals
Eukaryotes
Archaea
Bacteria
Prokaryotes
Origin of life
THE
TREE OF LIFE

APPENDIX II - PROTISTS

PROTISTS are either unicellular or multicellular organisms and are divided into two morphologically and ecologically distinct groups, PROTOZOA, which include animal-like protists, from which derive FUNGI and PROTOPHTA, which are plant-like protists, mostly ALGAE. Protists are evolution's first attempts at complex organisms, for they all have cells with nuclei, making them the basic building blocks for all animals, plants and fungi.

The kingdom Fungi comprises many species within various sub-kingdoms. A familiar sub-kingdom is Dikarya, which is divided into two phyla, Basidiomycota and Ascomycota. The former phylum contains all toadstools, mushrooms, brackets, puffballs and stinkhorns. The latter contains morels, truffles, baking and brewing yeast, and many lichens. The parts of fungi most evident are actually the fruiting bodies, while the living and growing parts of fungi are networks of threading hyphae called MYCELIA, which can be vast and ancient organisms hidden from view. Many are symbiotic with other organisms.

Sexual reproduction occurs in many fungi, which means that meiosis and fertilization provide the genetic variety for natural selection to evolve them more effectively in changing conditions. Simpler organisms generally don't require this genetic variety to survive because their chosen living environments have remained relatively more constant over the history of life. Mutations and prolific rates of non-sexual reproduction have been sufficient for them to cope with subtle changes that do occur.

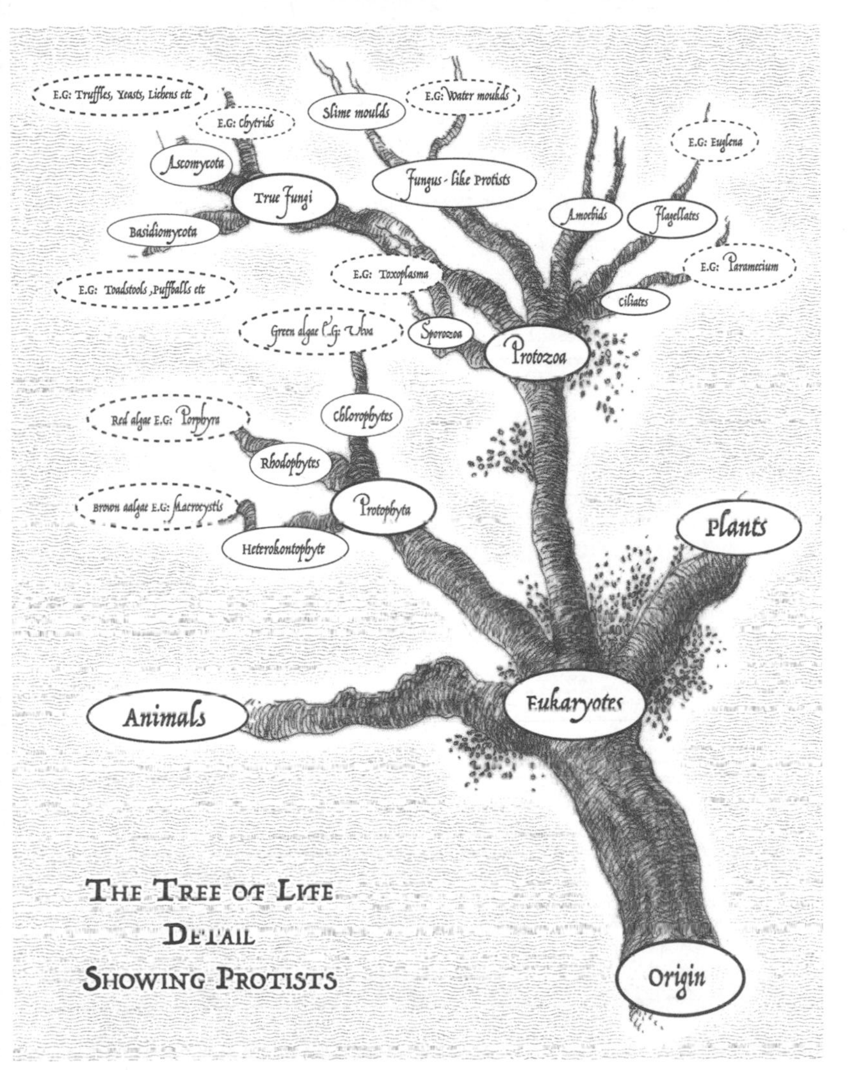

THE TREE OF LIFE
DETAIL
SHOWING PROTISTS

APPENDIX III - PLANTS

The plant and animal kingdoms exist in dynamic equilibrium, relying on one another for their survival. Plants absorb carbon dioxide and produce oxygen, while animals do the exact opposite. Plants provide food at the base of the food chain for animals, while animals return the nutrients to the soil by way of excreta and decomposition. Plants also bring new stocks of nutrients into the equation by deriving them from minerals, water and air. This interdependency between organisms is a vital characteristic of life on Earth, which is why the planet's ecosystem should be treated with respect.

PLANTAE, the plant kingdom, is divided primarily between VASCULAR and NON-VASCULAR plants. Vascular plants have the ability to transport fluids from roots to other parts of the organism, which means they can live on land and colonise areas where there is no surface water. Vascular plants include clubmosses, ferns, horsetails, conifers, cycads, ginkgos and flowering grasses, rushes, herbs, shrubs and all trees. There is a clear evolutionary progression from the production of spores to naked seeds and then to shelled seeds and nuts, which provide the germinating plant embryos with food and protection. Non-vascular plants are the most simple of land-dwelling plants. Generally small and shade-loving, they have no roots, stems, or leaves.

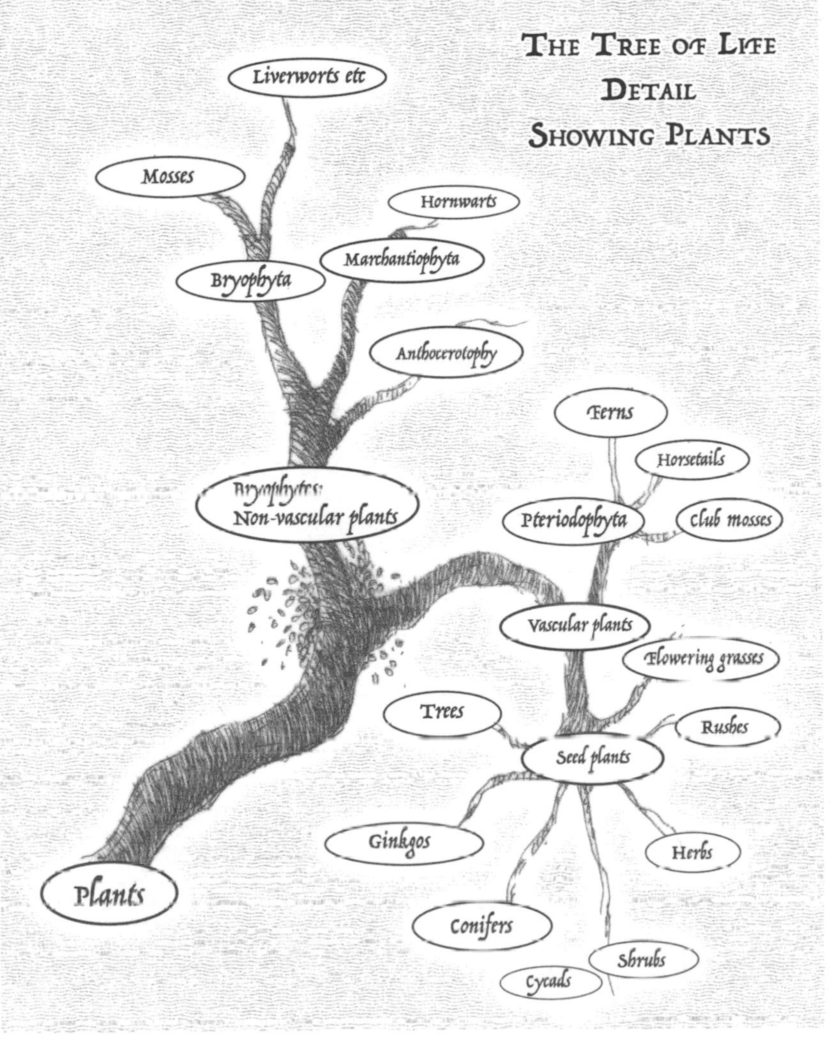
THE TREE OF LIFE
DETAIL
SHOWING PLANTS
Liverworts etc
Mosses
Hornwarts
Marchantiophyta
Bryophyta
Anthocerotophy
Ferns
Horsetails
Bryophytes:
Non-vascular plants
Pteriodophyta
club mosses
Vascular plants
Flowering grasses
Trees
Rushes
Seed plants
Ginkgos
Herbs
Plants
Conifers
Shrubs
Cycads

APPENDIX IV - ANIMALS

Animals range from simple unicellular organisms to highly complex multicellular organisms. They are motile, i.e., able to move spontaneously and independently (at least at some point in their lives) and they, like plants, consist of single cells or collections of cells that communicate and cooperate with one another. Most animal phyla appeared in the seas of the Cambrian era, around 550 million years ago.

The classical taxonomic division of the kingdom ANIMALIA is into VERTEBRATES and INVERTEBRATES (with or without spines). The invertebrate group comprises about 97% of all animal species and includes amoebas, hydras, sponges, worms, mollusks (slugs, snails), cnidarians (jellyfish, anemones, corals), echinoderms (urchins, starfish), cephalopods (squid, octopus, cuttlefish) and arthropods (crustaceans, arachnids, insects). The vertebrate group includes fish, amphibians, reptiles, birds, marsupial mammals and placental mammals. Each extant species is at the apex of its own evolutionary story.

The modern classification of the Animal kingdom involves 13 phyla, including at least three containing different kinds of worms. The largest phylum by far is ARTHROPODA, mostly populated by insects, with well over a million named species, and 20 million unnamed. In all, there are probably around 30 million species of plants and animals on Earth, of whom human activity is killing off about 50,000 a year, or 1% every 6 years, the fastest rate of genomicide since the K-T extinction which wiped out the dinosaurs along with 85% of all species on Earth. Last time it took 30 million years for the Earth to recover.

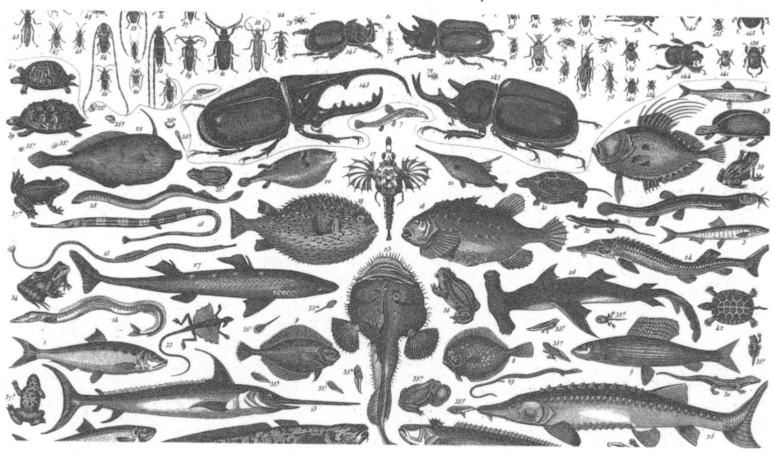

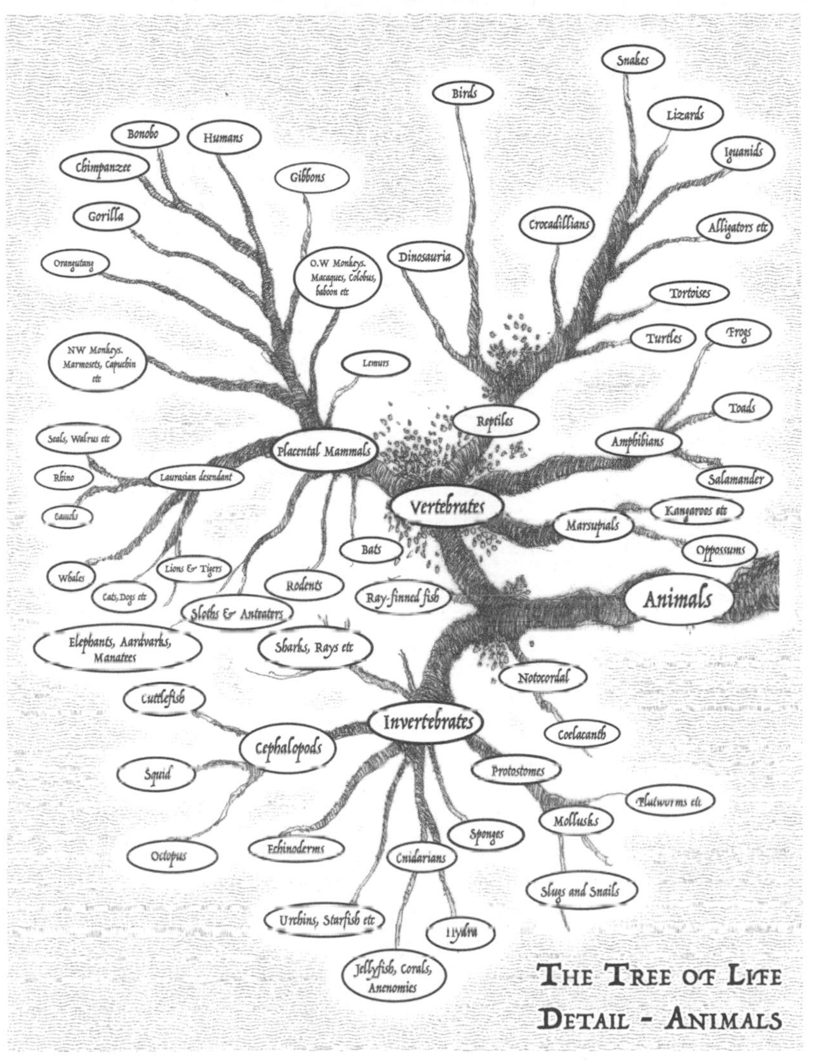

Snakes
Birds
Lizards
Bonobo
Humans
Iguanids
Chimpanzee
Gibbons
Gorilla
Crocadillians
Alligators etc
Orangutang
Dinosauria
O.W Monkeys. Macaques, Colobus, baboon etc
Tortoises
Turtles
Frogs
NW Monkeys. Marmosets, Capuchin etc
Lemurs
Toads
Reptiles
Seals, Walrus etc
Placental Mammals
Amphibians
Rhino
Laurasian desendant
Salamander
Vertebrates
Kangaroos etc
Camels
Marsupials
Bats
Oppossums
Whales
Lions & Tigers
Rodents
Cats, Dogs etc
Ray-finned fish
Animals
Sloths & Anteaters
Elephants, Aardvarks, Manatees
Sharks, Rays etc
Notocordal
Cuttlefish
Invertebrates
Coelacanth
Cephalopods
Squid
Protostomes
Flatworms etc
Mollusks
Sponges
Octopus
Echinoderms
Cnidarians
Slugs and Snails
Urchins, Starfish etc
Hydra
Jellyfish, Corals, Anenomies
THE TREE OF LIFE
DETAIL - ANIMALS

APPENDIX V - PHYLOGENY OF LIFE

Key: mya = million years ago
tya = thousand years ago
myf = million years in the future

Dates corresponds to the point at which each diverged from our collective common ancestor

2 myf: (Homo - ?),
300 tya-present: Homo sapiens,
300 tya: Homo neanderthalensis,
800 tya-300 tya: Homo heidelbergensis,
1.4 mya - 200 tya: Homo erectus,
1.9-1.5 mya: Homo ergaster,
2.4-1.9 mya: Homo rudolfensis,
2 mya: Chimpanzee & Bonobo diverge from their own common ancestor,
2.5-1.9 mya: Homo habilis,
3 mya: Australopithicus africanus,
3.9 mya: Australopithicus afarensis,
4 mya: Australopithicus anamensis,
5.8-4.4 mya: Ardipthicus ramidus,
6 mya: Orrorin tugenesis,
7 mya: Gorrilla,
7-6 mya: First Human-like species appears (e.g., Sahelanthropus tchadensis),
14 mya: Orangutans,
18 mya: Gibbons,
25 mya: Old World Monkeys (e.g., Macaques, Colobus, Baboons, etc),
40 mya: New World Monkeys (e.g., Capuchin, Marmosets, Spider Monkeys),
63 mya: Lemurs,
70 mya: Tree Shrews etc,
75 mya: Rodents & Rabbits (sharing their own common ancestor around 40 mya),
85 mya: Laurasian continent decendents (e.g., Cats, Dogs, Camels, Horses, Seals, Whales, Hippos, Bats etc),
80-105 mya: All other placental mammals: (e.g., Elephant, Manatees, Aardvark),
140 mya: Marsupials (e.g., Kangaroos, Opossums etc)
180 mya: Monotremes: Duck-billed Platypus.
300-220 mya: Reptiles and first true Birds. Turtles (300 mya) Crocodilians (240 mya), Snakes (220 mya) etc.
340 mya: Amphibians (e.g., Frogs, Toads, Salamanders etc),
415 mya: Lungfish
440 mya: Ray-finned fish (e.g., Herring, Salmon, Sturgeon etc)
460 mya: Sharks and Rays,
530 mya: Lampreys, Protostomes & Deuterstomes: (e.g., Flatworms, Velvet worms, Mollusks), Sea squirts,
1600-1000 mya: Sponges, Ctenophanes: Jelly-like organisms (e.g., Venus's Girdle), Cindarians (e.g., Jellyfish, Coral, Anenomes),
2500 - 1600 mya: Protists, Plants, Amoebas, Fungi,
3000.86 - 2500 mya: Eubacteria & Archaeans

APPENDIX VI - GLOSSARY

Adaptation. *The process of change by which an organism or species becomes better suited to its environment.*
Chromosome. *A thread of nucleic acids and proteins wrapped up in the nucleus of most living, organic cells. Chromosomes carry genetic information - genes.*
Crossover. *The shuffling of genes between homologous parental chromosomes during meiosis. Results in a gamete (sperm or egg).*
Ecosystem. *A biological community of interacting organisms and their physical environment.*
Environment. *The surroundings or conditions in which a person, animal, or plant lives or operates.*
Epigenetic Inheritance. *A term used in the study of environmental factors that influence the genome.*
Expression. *If a gene is active in a cell it is said to be expressing itself. The phenotype is an expression of the genome.*
Gene. *A stretch of DNA which codes for a protein.*
Gene Pool. *The complete set of unique alleles in a species.*
Genotype. *Describes the genetic constitution of an organism.*
Gamete. *Sperm or egg.*
Genome. *The DNA of a species.*
Lamarckism. *A theory that some acquired characteristics can be inherited by later generations. Operates via epigenetics.*
Meiosis. *A type of cell division where each of the two cells that result are a unique remix of parental DNA. Genes are shuffled between homologous chromosomes creating variety in the gametes.*
Meme. *A cultural equivalent of a gene.*
Memetics. *The study of memes.*
Mitosis. *Cell division where each of the two cells that result have equal number and kind of chromosome as the parent.*
Niches. *Describe (species) gaps in an ecosystem.*
Nucleus. *A safe place where DNA lives.*
Phenotype. *The outward manifestation of the (internal) genetic information of an organism. For example, blue eyes are the phenotype of the (genetic) information of an individual. Conversely, the trait for blue eyes is information held, but not always manifested, in the genotype.*